Intelligence and Development
A Cognitive Theory

Mike Anderson

BLACKWELL
Oxford UK & Cambridge USA

First published 1992

Blackwell Publishers
108 Cowley Road
Oxford OX4 1JF
UK

Three Cambridge Center
Cambridge, Massachusetts 02142
USA

British Library Cataloguing in Publication Data

A CIP catalogue record for this book is available from
the British Library.

Library of Congress Cataloging-in-Publication Data

Anderson, Mike.
Intelligence and development: a cognitive theory/Mike Anderson.
p. cm. – (Cognitive development)
Includes bibliographical references and index.
ISBN 0–631–16193–7. – ISBN 0–631–17455–9
1. Intellect. 2. Genetic Psychology. 3. Cognitive psychology.
I. Title. II. Series: Cognitive development (Oxford, England)
BF431.A5734 1992 153.9–dc20 91–44850 CIP

Typeset in 10 on 12pt Plantin
by Hope Services (Abingdon)Ltd.
Printed in Great Britain by T.J. Press Ltd, Padstow, Cornwall

Contents

Acknowledgements

This book has two origins. The first was in Edinburgh in 1976 when as a member of Chris Brand's differential psychology class I heard about newly published data on the relationship between inspection time (a measure of speed of processing) and IQ. The second was around 1984 at the Cognitive Development Unit. One morning at one of our seminars I suggested that an unexpected difference between two groups of subjects, matched for mental age, was due to the fact that one group had 'lower IQs' than the other. John Morton suggested that unless I could say what IQ meant in information processing terms I should desist from using such a prehistoric term. This book represents my best shot at satisfying John.

The approach I have taken in this book has evolved under the influence of many people. Chris Brand taught me about intelligence testing at a time when it was unfashionable in British psychology (sadly it still is). His dedication to scholarship is something I have long admired and respected. Chris has been an unwavering friend and I would like to think that this book repays some of the faith he has shown in me. As a postgraduate at Oxford I was lucky enough to have Pat Rabbitt as my supervisor. Pat taught me how to be a cognitive psychologist and above all else he showed me how exciting research could be. When I joined the Cognitive Development Unit Uta Frith encouraged my interest in intelligence at a time when few could see where the work might lead. Uta also introduced me to Neil O'Connor and Beate Hermelin whose great knowledge and experience have prevented me from continually reinventing the wheel. Most of all I am indebted to John Morton. He has nurtured the ideas in this book and, I am sure, brought out the best in them. He has been tireless in offering comments and suggestions and has pressed me long and hard to knock many ill-formed ideas into shape. His ability to detect ambiguities and his grasp of the major theoretical issues in cognitive psychology have been invaluable. More than anyone John has changed the way I think about intelligence and, indeed, psychology in general. He has yet to convince me, however, that Burnley Football Club will soon be challenging for the European cup.

Of course this book owes much to the other scientists, both past and present, of the Cognitive Development Unit. Weekly seminars shared with Rick Cromer, Uta Frith, Prajna Das Gupta, Mark Johnson, Annette Karmiloff-Smith and Alan Leslie provided a heady brew indeed. Many others who have spent time as visitors, students or research workers at the Unit, such as Jill Banks, Simon Baron-Cohen, Susan Carey, Mike Cole, Sybil Diskin, Suzanne Dziurawiec, Lindsay Evett, Anita Jackson, Hilary Johnson, Stephanie Keeble, Sigrid Lipka, Jean Mandler, Daphne Maurer and Dick Neisser, also left their mark. But the Unit was always more than a stimulating intellectual environment – it was also fun. I have great memories of Rick Cromer's hilarious lunchtime stories; of coffee, doughnuts and laughs shared with Anita, Elaine, Jill, Mani, Sigrid, and Suzanne; and of our various Christmas parties (Alan Leslie's rendition of 'My love is like a red, red rose', Warwick Smith's playing of the crumhorn synchronized with disco lights and Annette Karmiloff-Smith's characterization of Gloria Ramsbottom were but three of the highlights vying for top spot). Science is not conducted in a vacuum, and these enjoyable aspects of daily life, coupled with the responsibility of chairmanship, with Uncle Al, of the ethics committee and the incumbent tutelage of the young Markie-boy, have had a significant impact on the book.

The structure of the book emerged out of two courses of lectures I gave at the City University in London in 1987 and 1988. The Medical Research Council funded a trip to Australia in 1988 during which the first four chapters were drafted. My visits to different Australian departments were also funded in part by a Royal Society grant. The MRC continued to support me by way of a Senior Research Fellowship at Edinburgh University during 1989 and the first half of 1990. Chapters five, six and seven were drafted in this period, during which Chris Brand and his psychometric group kept me informed about intelligence, and the KGB (Stefana Broadbent, Francesco Cara, Jim Demetre and Tom Pitcairn), about development. Tom and Jim, in particular, were stimulating colleagues and we solved many psychological problems (and initiated many more!) over a beer and a curry. The final two chapters were drafted after my arrival at the University of Western Australia and the psychology department here has been more than generous in their support of my priority of finishing the book.

Thanks to Alison Mudditt for guiding the book through to production and to Ian Anderson, my nephew, and Maria Laura Speziali for their artwork.

The book has been much improved by the comments of colleagues on draft chapters. Both Annette Karmiloff-Smith and Suzanne Dziurawiec read all the chapters and provided very detailed comments. Annette's insights, particularly in the areas of modularity and development, fuelled me with new ideas and enthusiasm which carried me through the tedious final stages of revision, while Suzanne's radical surgery on my syntax and punctuation probably saved the proof reader's sanity. Dorothy Bishop, Uta Frith, Ted Nettelbeck, Neil O'Connor, Pat Rabbitt and Elaine Webster gave me feedback on the whole book while Chris Brand, Frank Dempster, David Hay and Beate Hermelin gave me feedback on individual chapters. Of course, the usual disclaimer applies – the final responsibility for the ideas expressed in this book remains with me.

Finally, I remember offering the opinion in 1987 that I could write this in six months. Instead the book has dominated my life for four years. Surviving this would have been a great deal more difficult were it not for the support of family and many good friends. My mum and dad valued learning and supported me through the university education denied to their own generation. My greatest sadness is that my dad did not live to see this book published, but I feel he would have been proud, as my mum is, of this most tangible manifestation of their own foresight. My wife Elaine has been through this thing with me from beginning to end. She has borne the late nights, lost weekends and obsessiveness (not to mention Dynamo Upholstery F.C.) with great grace and cheer. Friends who know us well, tell me I am a lucky man. They are right.

In memory of my father, Robert Edward Joseph
Anderson

1

The Scope of the Theory

This book is about intelligence *and* cognitive development. The choice of a conjunction is deliberate. There are many books about intelligence and many more about cognitive development. However, there are no books that are about intelligence *and* development. This is because intelligence and development are regarded as merely different ways of talking about the same thing. If we are interested in intelligence, we talk about the steady state structure of cognition; and if we are interested in development, we talk about how this structure changes. So, typically, psychologists theorize about intelligence, or they theorize about its development. This book is about intelligence and cognitive development because I think these terms refer to quite distinct cognitive properties. The former is not the steady state snapshot of the latter. To emphasize that this is not merely an exercise in semantics and to give you, the reader, a taste of the departure that this theory represents, I should tell you one of my main hypotheses:

Intelligence does not develop.

I am not referring here to the fact that a measure such as IQ shows some constancy over development. I mean something more significant than this. What I mean is that *individual differences in* and the *development of* intelligence are underlain by quite different cognitive processes. During the course of this book we shall see that there are many different kinds of views of intelligence, ranging from the biological to the cultural; but what they all have in common is the belief that, whatever it is that constitutes intelligence, it develops. By specifying the kinds of mechanisms that underlie intelligence and development, I hope to convince you that development can take place – that is, intellectual competence can increase – without a corresponding change in the mechanisms that are responsible for individual differences in intelligence. To put this endeavour in its proper context, let us first look at the range of alternative views on offer.

Low-Level and High-Level Views

It is no coincidence that the two broad areas of psychology currently vying for the authoritative position for theorizing about intelligence have their bases in two quite different camps: the study of adult individual differences and the study of cognitive development. The theories of intelligence inspired by these camps can be characterized as *low-level* and *high-level*, respectively. Low-level theorists regard intelligence as a biological, genetically determined attribute of our nervous system. High-level theorists, on the other hand, regard intelligence as a culturally determined, experientially driven attribute of *cognitive* functions.

Low-level theories are usually explicitly physiological, regarding intelligence as a parameter of neural functioning (Jensen 1982, Eysenck 1986). What distinguishes the intelligent from the unintelligent adult is synaptic efficiency, or speed. The strategy of low-level theorists has been to demonstrate that intelligence test scores can be predicted by performance on tasks which require little or no knowledge or by measures which are purely physiological. This concern with explaining differences in intelligence test scores as a function of differences in low-level information processing was originally inspired by the adult literature on individual differences. However, it has also been extrapolated, more implicitly than explicitly, to cognitive development (as we shall see in chapter 7). Since differences in test scores are the target of explanation, whether these represent differences between two adults or longitudinal changes within the same individual is seen as irrelevant. It is taken to be a parsimonious assumption that these differences in scores are to be explained with reference to the same mechanism. Thus, for example, higher synaptic efficiency makes one individual more intelligent than another, and increasing synaptic efficiency with age makes us more intelligent as we develop.

By contrast, high-level theories attempt to explain intelligence in terms of the cognitive mechanisms underlying human mental processes (Hunt 1980, Sternberg 1985). These cognitively orientated theories have been inspired by the developmental tradition which ascribes the increasing intellectual competence of the child to changes in the structure and use of knowledge (Piaget 1954). Cognitive theories view intelligence as a function of how knowledge that is available to the organism is used. Differences in intelligence among adults are based on differences in the knowledge base and the repertoire of strategic processes available for problem solving. As in mainstream cognitive development, the burden of explanation falls on elucidating the changing structure of knowledge with learning or development, explaining how these different structures are acquired, and, finally, understanding the processes by which they are used in intelligent activity.

There can be no more direct contrast between competing theories. For the neural-efficiency school, a theory of intelligence must be a theory of physiological processes, and knowledge is a mere epiphenomenon of these processes. For the cognitive school, intelligence *is* knowledge. However, both camps do agree on one thing: that individual differences in adult intelligence and developmental changes are to be explained with reference to the same mechanism. For the physiologist it is synaptic efficiency; for the cognitivist it is knowledge.

The theory presented in this book represents, then, a considerable departure from this position. It proposes that individual differences in intelligence are to be explained with reference to one set of mechanisms, and developmental changes in intellectual competence with reference to a quite different set. However, there is a final twist to our background tale. If there are two sets of mechanisms, should there not be two books, one about 'intelligence' and one about 'development'? This would miss the other central point of the theory:

Intelligence constrains development.

Although individual differences in intelligence and developmental changes in the intellectual competence of the child are based on quite different kinds of mechanisms, those mechanisms underlying individual differences *constrain* developmental change. The relationship between individual differences in intelligence and cognitive change will provide the major clues to the nature of the mechanisms underlying both versions of the construct 'intelligence'.

Resolving Conundrums

The theory set out in this book arose out of an attempt to synthesize what seemed like diverse and often incompatible data. The incompatibilities manifested themselves mainly as a set of conundrums. For example, if intelligence is a reflection of knowledge structures, as the cognitivists would have us believe, why is it that tasks that require merely a simple perceptual discrimination can predict how many words a child will know (Anderson 1986a)? This is easily answered if we consider intelligence to be a general property of our biology, as the neural-efficiency school would have us believe. However, if intelligence is a general property of our biology, why is it that some individuals of very low psychometric intelligence can perform remarkable cognitive feats such as, when given any date, calculating which day of the week it falls on (O'Connor and Hermelin 1984) or recognizing that a number is a prime number (Hermelin and O'Connor 1990)? Clearly, although intelligence and knowledge are related constructs, they are not synonymous.

A closer look at knowledge highlights another conundrum. Some kinds of computationally simple problems like mental arithmetic are usually too difficult for mentally retarded children, but they can perform other, almost wondrous, computational feats, such as extracting three-dimensional representations from a retinal input or an important syntactic distinction from speech. Our most powerful computers do not come close to matching even the mentally retarded on these kinds of computations; but on simple mental arithmetic the same children could not compete with your digital wrist-watch. There again, why is it that intelligence test performance predicts reading ability for the majority of children, but some very intelligent children are very bad at reading or spelling?

The kind of conundrums related above puts a premium on a theory of intelligence that can clarify the relationship between intelligence and knowledge. So, for example, we need to distinguish between two types of knowledge: that which relates to individual differences in psychometric intelligence and that which doesn't. Further, this imperative, to account for knowledge, constrains the kind of theory that can provide an adequate account of intelligence and development. Specifically, the theory will have to be a cognitive theory. That is, it will have to accommodate conceptions of the intellect that are expressed in theoretical constructs such as computation, representation, and information processing. However, it will differ from other cognitive theories of intelligence in that it will attempt to explicitly incorporate biological constraints on the acquisition and use of knowledge. In doing so, it will draw heavily on evidence and suggestions from a wide gamut of explorations of human intelligence, including those from *neurophysiological psychology, neuropsychology, psychometry,* and the emerging *cognitive sciences.* These disciplines take such radically different approaches to understanding intelligence and development that their theories remain blissfully ignorant of each other, often having dismissed the 'opposition' as irrelevant to their scientific enterprise. It is only by attempting a synthesis that the inherent contradictions, or paradoxes, appear. The success of the theory presented in this book should not be measured against each of its competitors *within* each of their own domains. Rather, the real test is how well the theory can bridge these different domains.

What Data must a Theory of Intelligence and Development Account for?

Successful scientific theories usually come about as a result of an attempt to explain something. In psychology, what it is that we would like to explain ranges from the grandiose (*the nature of mind*) to the pedantic (*the effect of caffeine intake on the self esteem of left-handed sophomores: a replication*). In the case of intelligence we are lucky. The study of intelligence can be as grandiose as we like. Intelligence is concerned with the mind: what we *know* and how we come to know it; it is about how we *think*, how we *reason* and *solve problems*; it is about the *rational* and *logical* side of human nature. Further, differences in human cognitive abilities are so significant and dependable as to render the study of intelligence both consequential and tractable. The technology of measuring such abilities, psychometrics, is primarily responsible for furnishing us with facts about intelligence of a reliability and importance unsurpassed in the whole of psychology. It is inevitable, then, that psychometrics should provide the bedrock of a theory of intelligence. In the next chapter we shall deal with the limitations and dangers of this tradition, but, for the moment, we must simply acknowledge its central role in providing most of our data. However, each of the facts listed below is relevant to a multiplicity of provocative data from other traditions, which justifies its inclusion as one of the central facts requiring explanation by our theory.

The data listed below constitute the core explanatory agenda of a theory of intelligence and development. It will quickly become obvious that the low-level and high-level views of intelligence have been able to survive this long in their splendid isolation only because each has its strengths in different parts of the data. The low-level view has its strength in being able to explain the *regularities* in the data; the high-level has its strength in being able to explain the exceptions to those same regularities. First, the regularities.

The regularities

(1) *Cognitive abilities increase with development.* This fact is consistent with any perspective on intelligence, but has been most adequately documented psychometrically. We can take a wide variety of cognitive tests, be they tests of verbal, spatial, mathematical, reasoning, or memory skills, in their myriad of forms, and we will find a clear developmental pattern. Children become more proficient at these tests as they become older, and they do so in a similar way for all the tests. That performance on cognitive tests from a variety of domains shows a very similar age profile, with a plateau around the late teens, is a robust and significant fact for any theory of intelligence.

This increase in cognitive abilities is indisputable, but the dramatic increase in scores during the developmental period obscures the fact that there are at least two quite different interpretations of the same score. It was Alfred Binet who first ranked intelligence test items on a difficulty dimension measured by the age at which a particular item was passed by a majority of children. This in turn yielded an interpretation of any particular score in terms of a *mental age* (MA). Thus, a child is given a mental age of 7 if the score obtained is typical of a 7-year-old, irrespective of how old the child actually is. William Stern (1912) had the insight that any particular score on an intelligence test can be given another interpretation (see figure 1.1).

Any score which can yield a mental age can also be interpreted as an *intelligence quotient* (IQ), which, in its original formulation, is:

$$IQ = MA/CA \times 100$$

IQ is, therefore, an indication of a child's performance relative to that of his or her age peers.[1] That the same score can be given two different interpretations is often overlooked today. For children, usually only mental age is considered, and IQ is ignored. This is because changes in cognitive abilities in development are so dramatic that the most obvious interpretation of any test score is in terms of what mental age the child has reached. In adults, where scores have, typically, stopped increasing, IQ seems the more obvious measure of intellectual functioning. However, IQ is usually regarded as being psychologically meaningless during development, being an artifact of the measurement of mental age. Therefore, mental age is regarded as being the appropriate mea-

---

[1]   This formula for computing IQ as a derivative of mental age is now an anachronism. Nowadays most tests are simply age-standardized, and IQs are derived from the normal distribution of scores.

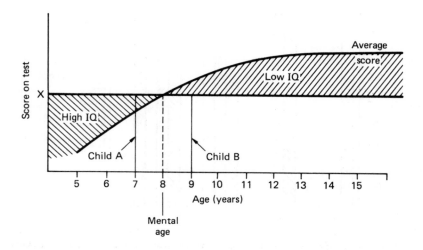

Figure 1.1 *The graph represents the average test score as a function of age. Both child A (7 years old) and child B (9 years old) achieve the same test score (X). This gives them the same mental age (8 years), but child A has a high IQ, while child B has a low IQ.*

sure of intelligence during development, because mental age reflects cognitive *achievement*. But regarding IQ as a meaningless statistical artifact does not gel with our second major empirical fact.

(2) *Individual differences are remarkably stable in development.* It is a fact that IQ measured at 5 years old predicts around 50 per cent of the variance in mathematics scores at 16 (Yule et al. 1982).

The year-to-year correlation between IQ scores is remarkably high, around 0.9, while over the whole period of schooling it is approximately 0.7 (Hindley and Owen 1978, see figure 1.2). Consider what this means. How well you are performing at 5 years old relative to other 5-year-olds predicts quite well how well you will perform at 16 years old relative to other 16-year-olds. Even more startling are new findings that report correlations between measures of cognitive processes in infants of a few months old and their IQs measured some years later (Fagan and McGrath 1981; Rose et al. 1986). This stability of individual differences, despite massive increases in intellectual performance (MA), suggests that IQ, rather than being a measurement artifact during development, is a psychologically real attribute of the cognitive functioning of children. Further, since the contents of knowledge systems

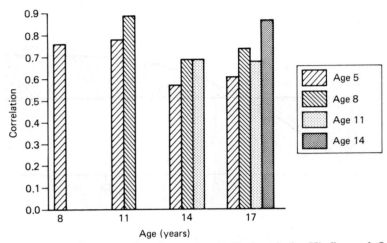

Figure 1.2 *The data are taken from a longitudinal study by Hindley and Owen (1978). They show that IQs derived from intelligence test performance are stable over ages. For example, the last block shows the correlation between the IQs of children when they were 5, 8, 11, and 14 years old with their IQ at 17. The correlations are greatest for the closest ages, but even at 5 the correlation with IQ at 17 is about 0.6.*

change so dramatically through development – and indeed, ability tests at 16 test very different knowledge domains than those at 5 (even more so at 5 months!) – this suggests that this attribute, whatever it is, cannot be equated with knowledge. This fact gels with our third key empirical finding.

(3) *Cognitive abilities co-vary.* When we examine our own cognitive abilities, it seems clear that we have strengths and weaknesses. We may think we are well read, but we are extremely poor at mathematics. We might be able to paint, but are hopeless at writing poetry. This all too easily leads us to imagine that intelligence is composed of isolated abilities. This is a pleasant thought. How good we are at one thing will not predict how good we are at another. Everybody has the chance to excel at something. Unfortunately it is not true that differences within individuals in cognitive abilities are at least as great as differences between individuals. If tests of a number of heterogeneous cognitive abilities are given to a random sample of the population, scores will co-vary. A child who is good at remembering lists of digits will also tend to know more words, will be better at deductive logic, and will be more proficient at reading. No one disputes this empirical fact, but the obvious conclusion, drawn by Charles Spearman (1904) at the beginning of the cen-

tury, that intelligence is general is most certainly contested. It has, in fact, been the object of a long and, as yet, unresolved debate. The theory advanced in this book proposes that the fact that cognitive abilities co-vary is a consequence of the cognitive architecture of individuals. In fact, it is the consequence of the same mechanism that underlies the stability in individual differences. General intelligence is not, as some have suggested, an artifact of test construction or the analytic techniques of psychometrics. Nor is it plausible that the co-variation of abilities is due to some non-cognitive factors such as motivation or socio-economic status. In a nutshell, general intelligence is a consequence of the constraint imposed by our cognitive architecture on our ability to think. The psychological reality of general intelligence has been the subject of many fierce debates, and we will turn our attention to this issue in the next chapter. Before we do so, we must introduce our final two facts, which may seem somewhat at odds with the previous three.

The exceptions

(4) *There are also specific cognitive abilities.* Although general intelligence accounts for most of the variation in intelligence test scores, there are other kinds of abilities which show individual differences. Psychometrically, these are usually called 'group factors'. Examples might be verbal ability and spatial ability. Although they have much in common (that is, performance on tests of these abilities co-varies), they also have something specific. Commonly we find that theories of intelligence propose *either* that intelligence is general (Eysenck 1986) or that it is composed of a collection of independent specific abilities (Gardner 1983). Those that countenance both (principally psychometric theories) simply state that there are both kinds of 'ability'. General ability has no special status as an ability, it is simply another one, albeit the 'largest'. Therefore, the psychometric tradition, although recognizing the distinction, sees no need to discuss their relationship. It must be a central issue for a theory of intelligence to explain the relationship between general and specific cognitive abilities.

A major question is, how many specific abilities are there? The strongest evidence, if we look at all our sources, is that there are two specific abilities. For example, cognitive psychology recognizes the same dichotomy between verbal and spatial abilities. They are thought of as depending on different modes of processing, and have been variously referred to as propositional and analogue, sequential and parallel, analytic and holistic, successive and simultaneous. While neurophysiology is happiest with the notion of general intelligence (synapses being common to all abilities) and has little to say about specific abilities, much neuropsychology is based on the very opposite idea, that of the localization of independent functions. For example, the notion that the left hemisphere underlies quite different abilities from those supported by the right hemisphere is as old as brain science. However, neuropsychology is unclear about what constitutes an ability, and as a result, much of its data which seems to argue for the existence of specific abilities actually argues for our last fact and our second major exception.

(5) *There are cognitive mechanisms that are universal for human beings and which show no individual differences.* I will argue that a failure to distinguish these kinds of mechanisms and those that underlie specific abilities is responsible for much of the confusion that surrounds the general intelligence/specific abilities debate. The mechanisms underlying specific abilities show individual differences. That is, we can be poor, quite good, excellent and so forth at these different abilities or modes of processing or hemispheric functions (as they may currently be described), and in ways not totally predictable from measures of 'general' intelligence. However, in addition, there are cognitive mechanisms that show no individual differences. We all see a three-dimensional world if we can see at all. We can all parse a speech stream into its syntactic components if we are not deaf or without language. Despite being of low IQ, children with Down's syndrome can understand that other people think and can have false beliefs, and act accordingly. Perhaps *only* some autistic people cannot do this (Baron-Cohen et al. 1985). All these kinds of abilities seem massively complex in comparison with the task of adding 2 plus 2; yet it is the latter, *computationally* simple task that children with Down's syndrome have problems with. It must be the case that the mechanism responsible for computing 2 plus 2, which is obviously functioning poorly in these children, cannot be the mechanism responsible for computing that 'because person A believes X, then it follows that A will exhibit behaviour Y'. The mechanisms responsible for these complex computational functions are, then, not specific abilities; they are universal *capabilities*. In cognitive science they would currently be thought of as modules. When they are damaged or missing, they become the object of interest for neuropsychology, in its search for the localization of psychological functions. Low spatial ability or low verbal ability can be thought of as a variety of low intelligence, but someone with a specific problem in understanding language would be thought of as *aphasic*, not unintelligent. A theory of intelligence and development must be able to explain why some knowl-

edge mechanisms are immune from the otherwise overwhelming imperialism of individual differences in intelligence.

The Plan

The agenda has been set; and although the foundations have only just been laid, already the rough outlines of the theory are taking shape. The theory will attempt to specify the mechanisms that underlie individual differences and their relationship with those that underlie cognitive development. Along the way it will have to elucidate the relationship between intelligence and knowledge, accommodating both biological and cognitive conceptions. In effect, by the end of the book we should have arrived at the specification of the *minimal cognitive architecture* that underlies intelligence and development and have shown how this architecture can accommodate the agenda laid out above.

Chapters 3, 4, and 5 will sketch in the different mechanisms in the architecture. Chapter 6 will discuss how these mechanisms influence cognitive development. Chapter 7 will tackle the hypothesis that, in one sense at least, intelligence does not develop. Chapter 8 will look more closely at the nature of mental retardation and the relationship between general intelligence and specific abilities in the light of the new theory. In the final chapter we will see how well the theory has measured up to the agenda and draw some general conclusions about the nature of intelligence and the implications for cognitive and biological theories of mind.

Before embarking on this enterprise, there is one major task left: we must rid ourselves of some of the mistakes, misconceptions, and ideological baggage (both well-intentioned and odious) that are the legacy of the often heated debate surrounding intelligence testing.

2

The Psychometrics of Intelligence: Fact or Fantasy?

The Political Issues

Sometimes I wish I were interested in mother–child interaction or some other *nice* area of developmental psychology. Being interested in intelligence so often carries unwanted baggage with it. The history of intelligence 'testing' is replete with gross misrepresentations of the data, shady scientific practice, and, frankly, frightening and spurious political analyses.[1] So, before I can begin to tell you what this book is about, I must first make plain what it is *not* about. It is not about differences in intelligence between blacks and whites, men and women, social classes, or humans and dolphins. It presents a theory of intelligence in terms of the cognitive architecture of individuals. Putative differences between groups are not within the scope of this theory. The theory will focus on what intelligence *is*, not who has more of it than whom. However, I cannot avoid the issue of the genesis of group differences altogether, because it seems that any theory may be pressed into the service of the cause of establishing that supposed racial differences are genetic in origin, no matter how inappropriate this may be. A theory that emphasizes biological contributions to individual differences in intelligence has a fair chance, judging by the historical record, of being misinterpreted on this most sensitive of issues. Even as I write, a new, and perhaps even more invidious, genetic theory of race differences has rekindled the controversy; and, unsurprisingly, it has as a central plank the claim that there are genetic differences between the races in intelligence. It is incumbent on me, therefore, to state clearly

[1] Stephen Jay Gould's *The Mismeasure of Man* (1981) documents the notorious history of the many attempts by political fanatics, masquerading as scientists, to 'measure' intelligence and subsequently draw prejudiced, and spurious, conclusions about the structure of society.

where the conception of intelligence advocated in this book stands with respect to these issues.

The hypothesis that the races may differ in intelligence pre-dates modern psychology, but was revamped in scientific terms in the late 1960s. In 1969 Jensen published a seminal article which put forward the hypothesis that *part* of the reason why black Americans do not fare so well in school may be because there are genetic differences in intelligence between the races. This caused a furore both in academia and in a wider public forum. The turbulence caused by this issue probably set back the scientific study of individual differences in intelligence for at least fifteen years, and certainly halted the progress of human behaviour genetics, even though very few behaviour geneticists studied racial differences (Plomin 1990). It is distressing, therefore, to see the issue raise its head once more and in a potentially even more explosive form (Rushton 1988, 1990; and see Flynn 1989, 1990, and Zuckerman and Brody 1988 for rejoinders). I will discuss Rushton's views (albeit briefly), not because they are worthy of serious consideration, but because they contain a number of important lessons for those who worry about the relationship between science and social responsibility.

It will be of no surprise to students of the history of race differences research to learn from Rushton that black Americans seem to fare rather poorly on many socio-economic variables (for example, crime rates, marital stability, school performance, and mental health) in comparison with white Americans and Orientals and that this is to be explained with reference to the different gene pools available to each racial grouping. However, the new version of this old story does contain a few novel features, features which should serve to heighten the sense of absurdity and which would turn the whole episode into a farce were the implications not so serious. The novelty is in matching data on physical features and sexual behaviour allegedly associated with racial groupings with a perfectly respectable theory taken from evolutionary biology: r/K selection theory (see Wilson 1975). This theory posits a continuum of reproductive strategies which expresses the degree to which parental care and the production of offspring are traded off. An extreme r strategy involves maximizing the number of offspring by sacrificing all parental care: *survival by numbers*. The r strategy is typified by oysters, who produce millions of offspring per year. An extreme K strategy involves maintaining a low birth rate but ensuring that a high proportion of offspring survive by maximizing parental care: *survival by nurture*. The K strategy is typified by the great apes who produce one offspring every two or three years.

Rushton assembles a mass of data which allegedly substantiate his hypothesis that 'Orientals' are most K-selected and 'Negroids' the least, with 'Caucasians' somewhere in the middle. Some of the data have antique status in the racial differences debate (for example, cranial capacity, brain weight at autopsy) and has been discredited before (Gould 1981). Some data constitute new collections of measures of sexual behaviour and characteristics (for example, sperm counts, size and position of genitalia, length of pregnancy, age of first intercourse, frequency of intercourse, frequency of sexually transmitted diseases) and some data are presentations of social statistics (for example, crime rates, illegitimate births, divorce, child abuse, and mental health). Rushton claims there is a remarkable pattern or rank ordering which places Caucasians between Orientals and Negroids on all these diverse variables. Further, the pattern of data can be made sense of by linking them with r/K selection theory. That is, Negroids have been more r selected (greater effort on producing offspring than caring for them) than either Caucasians or Orientals. This means that Negroids will produce more children and care for them less, resulting in the disintegration of social cohesion. Orientals will have fewer offspring, but will invest far more in parental care. This particular strategy is thought to require more intelligence, thereby explaining why, as a group, Orientals are more intelligent than Negroids.

Although it is not my intention to give credence to such a specious argument by taking issue with Rushton (and it would take us too far from the purpose of this book), it is incumbent on us to take note, at least, of the major objections to Rushton's work put forward by Zuckerman and Brody (1988), lest anyone be lulled into suspending disbelief:

1   The major empirical and theoretical link between the diverse data sets is simply not there. There is no evidence that these physical and behavioural characteristics are correlated with fertility.
2   Many of the sources of data are barely credible.[2]
3   The data are selectively reported.
4   Most of the comparisons do not control for obvious differences in social class. Clearly, the most obvious interpretation of what reliable data there are is that any differences between races are explained

[2] The sources of the data on sexual behaviour and characteristics are not always what we might expect from scientific presentations. Zuckerman and Brody (1988) point out that the reports of a French colonial army surgeon from 1898, cited by Rushton, on the placement of female genitals and angle and texture of erections is 'hardly a credible and unbiased source of anthropology' (p. 1026).

not by differences in genes but by differences in socio-economic circumstances which, unfortunately, are correlated with race in our
Western society.

More specifically, Brody takes particular issue with Rushton's claims
about intelligence in his theory. Two of the criticisms are damning.
First, although there can be little doubt that intelligence, as measured
by intelligence tests, is a heritable trait (for example, see Plomin and
Daniels 1987), evidence for within-group heritability does not constitute evidence for a genetic basis for between-group differences. Genetic
and environmental differences are necessarily confounded in any
between-group comparison. Second, there is little evidence, in any
case, that fertility is heritable or that intelligence and fertility are correlated once socio-economic class is controlled for.

The important lesson that can be learned from this most recent
attempt to give a genetic account of race differences is that it shows
that any theory (not just one about the nature of intelligence) can be
used inappropriately in the service of this particular cause. This time
the hijacked theory comes from evolutionary biology; in the past such
theories have been psychological theories (particularly of intelligence).
r/K selection theory was developed to explain the evolutionary strategies of species as different as oysters, dandelions, flies, rabbits, mice,
and men. Oysters produce about five million eggs a year, whereas the
great apes produce about one infant every five years. Clearly, on this
theory the *whole* of the human species is at the K or nurture end of the
r/K continuum. Yet Rushton proposes that we can apply this between-
species dimension, where offspring numbers range on a scale from one
to millions, to group comparisons *within* a species (human) in which
the group means of number of offspring must differ in tiny fractions per
year, if at all. Quite apart from the many inconsistencies in the logic of
the theory, as used by Rushton, and the arbitrary manner in which
characteristics are supposed to be classifiable as indexical of one strategy or another, the application of this theory in circumstances in which
its central assumptions (a reasonable range in reproduction rates) are so
obviously violated makes a nonsense (albeit a dangerous nonsense) of
the whole enterprise. In Rushton's case the mistake is to take a theory
from evolutionary biology that was designed to account for one set of
data (between-species differences) and apply it to an inappropriate set
(supposed racial group differences within the human species). In the
same way, it is a mistake to take a theory of individual psychology (such
as a theory of the nature of individual differences in intelligence) and
draw inferences about the cause of differences in an inappropriate data

set (between-group differences in intelligence test performance). There is simply no necessary connection between causal factors in individual differences and causal factors in group differences.

The theory advocated in this book will emphasize the importance of biological constraints on intelligence. I hope that the above discussion will keep this theory from falling victim to the fate that befell r/K selection theory. In case any doubt remains, let me be clear. If it turns out to be true that our biology constrains how we think, this provides no grounds for believing that any differences observed between groups on intelligence tests must also be due to differing biological constraints between the groups. The biological constraints in the theory apply to technical notions of how brains implement thinking. *Groups don't think.* The theory of biological constraints cannot, then, be applied to any putative group differences. The theory should not, therefore, be used to justify views that race or social class differences are *likely* to be due to group differences on the biological variable that is causally related to individual differences. There are no grounds for believing so. On the contrary, given that the theory also recognizes environmental (information input) constraints on the development of thought, and given that there is a great deal more evidence for environmental and cultural differences between races and social classes than there is for genetic differences, any between-group differences found are more likely to be cultural in origin.

The fact that I propose that individual differences in, and the development of, human intelligence are constrained by biology should have no political implications in any case. Nor should such issues, while of obvious social concern, act as spurs to scientific inquiry. We must be clear why this is so before we return to the substantive issue of the status of the psychometric evidence for individual differences.

First, most of the political debate centres on the well-known group differences in *intelligence test* performance; so we must be clear about the status of intelligence tests for this theory. A theory of intelligence is a theory of thinking. Although intelligence tests are the best measures of human thinking that we have, this does not commit us to believing that current intelligence tests are good measures of the theoretical construct 'intelligence'. Intelligence tests are imperfect instruments. Differences between groups on intelligence tests could be due to many factors completely unrelated to the biological constraint on the ability of individuals to think that is postulated by the theory.

Second, the most fundamental reason why there should be no political implications of this theory is that scientific facts or theories, while useful for countering incorrect assumptions about the nature of the universe, do not circumvent the process of political choice. For example, while it is a scientific 'fact' that black Americans as a group score around 15 IQ points lower than white Americans, at least two political options are left open: I could conclude that because of this fact blacks should be denied access to higher education, be paid less, and have poorer housing; but I could just as easily conclude that blacks should be given more education, more money, and better housing. The choice is a political choice. Science can never – and should never – be allowed to choose for us. The choice should be a question of morality, not science.

Finally, we may wonder what purpose such group comparisons serve. While it may be the case that comparisons between races or classes on a number of measures may constitute a necessary test of a sociological theory, it is my view that such comparisons add nothing to our understanding of individual differences in intelligence, at least in so far as this is a pursuit of a scientific psychology. Differences between tall and short people, between people who like dogs and those who like cats, between people who play golf and those who go swimming, are no more, and no less, interesting for such a theory. Quite simply, these differences are irrelevant for the important theoretical constructs which relate to individual psychology. If there is such a thing as the intelligence of a group it is a completely different construct from the intelligence of an individual, and it is the latter which is the concern of this book. This certainly means that racial and social class comparisons cannot be defended on the grounds that they advance our scientific understanding of 'human intelligence', as the term is usually understood. On this matter I advocate what I call 'Chomsky's dictum': 'Does the research in question carry costs, and if so are they outweighed by its significance?' (Peck 1987, p.200) In the case of race and IQ:

> It is difficult to be precise about questions of scientific merit. Roughly, an inquiry has scientific merit if its results might bear on some general principles of science . . . An inquiry into race and IQ, regardless of its outcome, will have a social cost in a racist society . . . The scientist who undertakes this inquiry must therefore show that its significance is so great as to outweigh its cost.(Peck 1987, pp. 200–1)

Since for a theory of individual differences in intelligence, there is no scientific merit in investigating race differences in IQ, we should devote no more space to the issue.

The Legacy of the Race Difference Debate

Aside from the obvious political issues, this debate has had important implications for anyone interested in theories of intelligence. The legacy of this issue has been an unwillingness, on the part of many psychologists, to believe that there is such a thing as intelligence. The historical precedent for this view is quite clear. In an attempt to refute the hypothesis that race differences in intelligence tests reflect underlying genetic differences between races, the tendency has been to deny that intelligence tests measure intelligence.[3] Perhaps 'intelligence', as a construct illuminated by intelligence test data, is an invention, rather than a discovery of psychometrics. There are two important arguments that have evolved from this initial reaction against the validity of intelligence testing: the first suggests that intelligence is a tautology, a function simply of the way intelligence test items are selected; the second argues that even if the selection of items were justified on *a priori* grounds, the abstraction of the construct 'intelligence' from test data is simply a mathematical abstraction with no concordant psychological reality. I will deal with each of those issues in turn.

Intelligence as a Tautology

People vary in performance on any task. If by intelligence we simply mean some kind of average of all the possible test scores that an individual could obtain on any task, then the notion would be uncontentious but meaningless. An individual's intelligence would vary dramatically with the kinds of tasks used. My average score from tests of macramé, chess, and cricket is likely to be very different from my average score from tests of computer programming, monopoly, and football. The construct 'intelligence' is only meaningful – that is, it can only be construed as some kind of cognitive capacity *underlying* the performance we measure – if it shows some stability across heterogeneous

---

[3] I think that this is a dangerous line of argument. It is as if the racist case would be justified if it turned out to be true that blacks were genetically less intelligent than whites. It is better to defend the moral stance that any such difference should be, in any case, irrelevant for social policy. Again Chomsky puts this well: 'they may enter the arena of argument and counter argument, thus impicitly reinforcing the belief that it makes a difference how the research comes out, and thus tacitly supporting the racist assumption on which this belief ultimately rests. Inevitably, then, by refuting alleged correlations between race and IQ . . . one is reinforcing racist assumptions' (Peck 1987, p. 201).

tasks that we would consider to require intelligence. Therein lies the tautology. What are these tasks? Intelligence is given meaning only by restricting its usage to particular types of abilities. But which abilities? Many of the critics see these abilities as being chosen, at best, arbitrarily, and at worst, to sustain a particular social system which discriminates against the poor, the uneducated, and ethnic minorities. What governs whether a test will be considered to require intelligence? One strong criterion is that performance on any new test should correlate with performance on the old ones, but this seems merely to perpetuate the tautology. Is our current notion of intelligence locked for ever in a self-perpetuating circle originating in an arbitrary, or politically motivated, selection of test items?

The Origin of Test Items

Intelligence tests were originally designed in the tradition of Binet, who sought to discover the important dimensions of individual differences in mental abilities by empirical means.

> We cannot wait for scientific studies to indicate the more important processes. . . . We must search with the present knowledge and methods at hand for a series of tests to apply to an individual in order to distinguish him from others and to enable us to deduce general conclusions relative to certain of his habits and faculties.[4]

This started psychometrics off on an empirical endeavour which it has pursued ever since. But was it only brute empiricism that was to determine what the important dimensions of human differences are?

> The goal pursued is not to determine *all* the differences in psychic faculties . . . but to determine the more marked and important differences. The tests must show the characteristic traits that distinguish the two individuals, not all their traits. That is a rule other authors have not considered and followed, for, if they had been clearly aware of it, they would not have set up so many tests for determining the *most elementary sensations and processes*. We must, therefore, deal with the most superior faculties. (Emphasis added)

---

[4] This and the translations which follow are from Binet and Henri and are taken from R. J. Herrnstein and E. G. Boring (eds), *A Source Book in the History of Psychology* (Cambridge, Mass.: Harvard University Press, 1965).

The empiricism of the early psychometricians was not, then, completely blind. They started off with an intuitive notion of what counted as a superior faculty. For Binet, the superior faculties were:

> memory, the nature of mental images, imagination, attention, the faculty of comprehension, suggestibility, aesthetic feeling, moral feelings, muscular strength and will power, and motor skill and perceptual skill in spatial relations.

So the selection of items for the first intelligence tests was not arbitrary. It was guided by what were considered at that time to be the important faculties of the mind. It is interesting that although there are many who do not see the current content of intelligence tests as a test of 'true' intelligence, they too have their own intuitive concepts of what is meant by intelligence. For example, Gardner (1983) regards symbolic processes as among the hallmarks of intelligence; for Gould (1981) it is the ability to deal flexibly, or creatively, with problems (this is a very commonly held intuitive theory). Clearly, a notion of intelligence that is meaningful must exclude *something*. The legacy of the Binet tradition is that thinking is regarded as central to most notions of intelligence. The list of thinking abilities that constitute a modern intelligence test is truly prodigious. The British Abilities Scales (Elliott 1983), for example, consists of 23 subscales, ranging from the measurement of *speed of information processing* (quickly cross out the circle with the most squares in it) to *social reasoning* (the child is presented with a real life, moral dilemma and is required to reason about it). Even if we consider that intelligence tests measure only a portion of cognitive functioning it is clear that (1) that portion contains a wide, non-trivial range of abilities, and (2) that we are hard pushed to think of any significant abilities which would be considered to be a subclass of thinking that has been left out.

What distinguishes Binet and the psychometricians from some modern theoreticians of intelligence is that, once they had settled on what were considered to be the important faculties (a theoretically motivated decision), empirical criteria became of paramount importance in shaping future intelligence tests. If a new test displayed the same properties as the old – that is, performance improved with age – and if, by and large, performance on the new test showed some relationship to performance on other tests, then it could qualify as a test of some aspect of intelligence. This process led to the development of new tests and the discarding of some redundant old ones. It also provided a method of empirically delimiting the concept of intelligence. Any old suggestion of what kind of activity required intelligence was no longer good enough; it must pass empirical muster. There was, of course, a price to be paid for this. For too long, psychometrics neglected serious theoretical work. Instead, it relied on the method of factor analysis to 'discover' the underlying properties of the mind that generated the pattern of results found by their battery of tests. This method, as we shall see, was no substitute for the experimental test of a genuine theory of intelligence, and led to sterility in debate and decline of interest in the field.

It is not without irony that the content of the early tests was determined not only by theoretical considerations about the nature of intelligence, but also by the attempt to *minimize* the importance of social and educational factors on performance. Rather than being motivated to develop tests that would favour the educated middle classes, Binet, on the contrary, tried to develop tests that would tap underlying intelligence rather than educational experience. As is well known, Binet was commissioned in 1904 by French authorities to devise tests which would identify children in French schools who required special education to improve their poor performance. Binet set out to remove the disadvantages that less well educated children might have on a test, simply because of their decreased opportunity to acquire certain kinds of knowledge. The idea was to allow some poor children to demonstrate that their substandard performance in school was due to a lack of educational opportunity rather than any inbred deficiency in intelligence. Intelligence tests were originally designed, then, to play down the possible effects of practice and experience within specific knowledge domains. This is the principal distinction between an intelligence test and a school exam.[5] However, many would like to see a kind of positive discrimination in these modern times. If we are interested in measuring intelligence we have to test those activities that people actually do, like betting on racehorses (Ceci and Liker 1986), rather than some obscure paper and pencil activity unrelated to real life. But this approach runs the danger that Binet attempted to avoid: that performance on the test would necessarily confound differential experience and aptitude. In fact, this approach *assumes* that intelligence is nothing other than an

[55] The fact that many modern intelligence tests contain scales which measure general knowledge does not reflect a change from the benevolent attitude of Binet so much as the empirical fact that in our society performance on these kinds of tests correlates with performance on others that do not reflect acquired knowledge. This 'fact' will become important for the theory advocated in this book, but for the moment let me simply note that if all subtests of 'general knowledge' were expunged from the testing repertoire, it would not change any significant conclusion that I would draw about intelligence from the psychometric data.

arbitrary collection of skills. We will see below that this is inconsistent with the psychometric data.

We can conclude that the nature of modern intelligence tests has been determined by, admittedly vague, theoretical notions as to the nature of intelligence and by empirical trial and error. The empirically based construction of tests is guided by the axiom that they must meet certain validity criteria, such as performance increasing with age and correlating with other kinds of tests. The latter criterion also has an unexpected consequence. Many of the proposed 'new intelligence tests', perhaps the most celebrated being tests of creativity, have simply not emerged, because it was quickly realized that they were so highly correlated with older intelligence tests that they added nothing new. There has been a conspicuous failure to discover tests of cognitive abilities that do not correlate with already established tests of intelligence. Seeking tests of new abilities which must correlate with older tests to qualify as measures of 'intelligence' does not lead *inevitably* to a tautology. It is an interesting empirical fact, on the other hand, that we can find many diverse cognitive tasks that are related to each other. It need not be so. Indeed, this fact is central to the whole notion of intelligence and, because of this, has been the focal point of argument about the psychological reality of the construct.

Intelligence as a Statistical Artifact

There have been two theories that have dominated psychometric conceptions of intelligence. The first views intelligence as consisting primarily of a single general ability (Spearman 1904). The second takes the contrary position that intelligence consists of a small number of independent abilities (Thurstone 1938). Both these theories are based on the same data set – namely, the pattern of intercorrelations found in performance on a variety of intelligence tests. The nature of this data is not subject to dispute. When a heterogeneous set of cognitive tests – for example, tests of imagery, memory calculation, vocabulary, and so on – are given to a random sample of the population, scores on the tests will correlate. The correlation matrix is said to have a *positive manifold.* This means that the tests have something in common.

It was Spearman's statistical mastery of the correlation coefficient that led to the development of factor analysis. Basically, what factor analysis does is to take a data matrix of intercorrelations and search for underlying traits that could have generated the patterns of co-variance seen in the data matrix. The problem is that for any one data matrix there are a large number of possible solutions; that is, there are many combinations of possible underlying factors or traits that could have generated the matrix. Taking the view that the factors extracted should themselves be unrelated to each other, Spearman claimed that the method of factor analysis revealed one large factor accounting for most of the variance and a number of small additional factors specific to each test. Spearman called this large factor 'general intelligence', or g.

Since Spearman, British psychometricians, while acknowledging the 'dominance' of the g factor, have also recognized that other factors account for significant proportions of the variance in intelligence test scores. This view, that intelligence consists of a hierarchy of abilities with g at the base, was best articulated by P. E. Vernon's (1950) theory of the structure of the intellect. Vernon emphasized that when g is extracted from a matrix of correlations generated by a heterogeneous set of cognitive tests, the remaining variance is not specific to each individual test, as Spearman supposed, but rather can be allocated to two main groups of tests. The major *group factors* described by Vernon are v:ed, or verbal educational abilities, and k:m, spatial mechanical abilities. These groupings can be thought of as indicative of more specific abilities, one level up from g in the psychometric hierarchy of abilities.

On the other side of the Atlantic, Thurstone saw no particular merit in the requirement that the extracted factors be uncorrelated, preferring instead to select factors which corresponded to a *simple structure* among the test intercorrelations, where the factors extracted correspond to clusters, or groupings, of tests. These factors were themselves correlated, but this loss of analytic power was considered a small price to pay for salvaging the plausibility of the factors as psychological faculties. Because Thurstone's factors corresponded, loosely, to clusters of tests, whereas Spearman's g was necessarily common to all the tests, this makes it easy to 'see' what Thurstone's factors 'are'. (This, in fact, is all that constitutes psychological plausibility!) He identified eight factors in all, giving them the name 'Primary Mental Abilities'. These were verbal, word fluency, number, spatial visualization, rote memory, inductive reasoning, deductive reasoning, and perceptual speed. Like Spearman in Britain, so Thurstone set the tone in the United States, with American psychometrists (the most obvious example being Guilford 1966) tending to play down (or simply ignore) the importance of g and play up the independence of abilities.

In sum, the method of factor analysis, the cornerstone of the psychometric tradition, had not lived up to its early promise as an objective method of determining the nature of human intelligence. What we find, instead, is the peculiar (some may say damning) situation of the same data matrix, containing the undeniable co-variation of abilities, being analysed in different ways to support diametrically opposed theories. Psychometricians have tried to find objective methods to find the best solution, such as *simple structure, principal components,* or whatever. We should not worry about the rationale or detail of these solutions. It is enough to know that at the end of the day their choice may seem arbitrary but is, in fact, driven by the psychometrists' prior 'theory' of the structure of intelligence. A principal components analysis will generate a solution to a matrix that will yield a large general component accounting for something like 40–50 per cent of the variance, with smaller 'group' components subsequently being extracted. This produces a solution that emphasizes general intelligence. On the other hand, a factor analysis using the method of simple structure can rotate out the variance found in our first principal component and disperse it among higher-order factors. We see, then, that two different ways of analysing the same data matrix can produce two very different pictures of the underlying abilities that putatively generated the matrix in the first place. So changing the method of analysis changes the theory.

This has reinforced the view of those who believe that the whole notion of intelligence is an artifact of the psychometric method. Such a fundamental theoretical difference as that between general intelligence, or *g*, and a multiple intelligence position turns out, they claim, to be nothing more than a dispute over mere mathematical abstractions arbitrarily chosen from an infinite supply. As Gould (1981) has put it:

> Factor analysis is a fine descriptive tool; I do not think it will uncover the elusive (and illusory) factors, or vectors, of mind. Thurstone dethroned *g* not by being right with his alternate system, but by being equally wrong – and thus exposing the methodological errors of the whole enterprise. (Gould 1981, p. 310)

Despite Gould's apparent even-handedness, it is the notion of *g*, or general intelligence, that has come in for particular vilification. This is because the status of general intelligence as a cognitive ability seems inextricably linked to psychometric techniques, whereas Thurstone's Primary Mental Abilities have a surface validity as *psychological* constructs. While it must be the case that psychological plausibility should guide the psychometrist in his model-building, psychological plausibility boils down to having a theory. One of the major goals of this book is to present a theory which makes plausible the psychological reality of general intelligence. The motivation for doing so is founded on the view that general intelligence is real and that, by contrast, notions of multiple independent intelligences, or faculties, are the artifactual products of statistical machinations.

There are three principal reasons for believing that general intelligence is real.

1   Even within the psychometric framework it is not true that the data are as ambiguous as is usually claimed. More modern analytic techniques such as LISREL (Joreskog and Sorbom 1981) have shown that although there is good support for group factors (such as those proposed by Burt and Vernon) and for more specific factors (roughly equivalent to the Primary Mental Abilities of Thurstone), *hierarchical* models with a unitary *g* factor at the apex of the hierarchy fit the data better than any model without such a *g* factor (Gustafson 1984). The distinction between fluid and crystallized *g* made by Cattell provides an interesting illustration of this point. This is the American model that comes closest to conceding that Spearman's *g* is 'real'. Fluid *g* emerges from an analysis of tests involving novel problem solving, such as *analogies, matrices,* and *series completion,* while crystallized *g* emerges from an analysis of tests reflecting factual knowledge and verbal and numerical skills. These are considered to be independent abilities in the Cattell–Horn formulation (see Horn and Cattell 1966). However, they are still correlated with each other, and the hierarchical analysis shows that the Cattell–Horn fluid intelligence factor is *exactly* equivalent to Spearman's *g* factor. Certainly Gould (1981) is right to claim that factor analysis merely affords description: alternative factor analytic solutions do not generate alternative *theoretical* structures; they simply generate alternative descriptive or terminological structures. But what is clear is that any theory of intelligence *must* incorporate a picture of the intellect in which the same general component appears in all measures of thinking. This component may be *hidden* by the judicious choice of factor-analytic technique, but it will always reappear as a set of intercorrelations among the factors among which the *g* variance has been dispersed. Contrary to Gould's claim, *g* cannot be made to disappear by selecting an alternative method of analysis; only an incomplete analysis can do this.

2   The corollary of this argument is that we can design tests that measure general intelligence and very little else, but no one has been able to construct a test battery to measure a collection of independent abilities where no general factor can be extracted.[6] It is not as if it has

---

[6] Raven's progressive matrices, for example, are considered by some to be measures of *g* and very little else.

not been tried. The logical conclusion for Thurstone after 'discovering' the Primary Mental Abilities (PMA) was to design tests that would measure only a single ability. Unfortunately this project failed. Nearly all the tests specifically designed to test an independent PMA had higher loadings on g than they did on the PMA factor in question (Eysenck 1939). Similarly, the British Ability Scales, which represent the state of the art in intelligence test construction, were designed to measure homogeneous components of cognitive ability:

> It was decided that the factor analytic approach should be abandoned as the main technical basis of the scales . . . Instead, it was decided that the Rasch model of item analysis would be used. . . This carried the immediate implication, since the Rasch model is unidimensional, that the scales which were finally produced should be homogeneous in their item content. This seemed very much more in accord with the developments in professional practice . . . with a reduction in emphasis on IQ scores and an increased emphasis on the need for information on cognitive processes. (Elliott 1983, p. 24)

In other words, the scales (all 23 of them) were each constructed to measure a different facet of cognition, and were intended to measure only that facet. However, when administered, they, in common with all intelligence tests before them, yield a principal components solution with a general factor being the largest single factor in the test battery (Elliott 1986). If any battery of tests was ever likely to yield the multiple independent intelligences solution simulated above, it was this one. The fact that it does not do so must surely convince us that, rather than general intelligence being an artifact of statistics, it is multiple intelligences that are the illusory creation of statistical sophistication.

3   The mistake (as some see it) of reifying the positive co-variation of abilities as an attribute of the real world and, more particularly, as an attribute of the cognitive architecture of individuals is based on the failure to distinguish between a mathematical abstraction obtained from population statistics and a psychological or biological mechanism. This, indeed, is a failure of much psychometric theorizing. Let me stress that a population statistic and a psychological function represent different levels of description and explanation. However, whether the co-variation of abilities should be reified as a psychological mechanism depends on there being a *theory* which attributes the former to the operation of the latter. The fact that there is a distinction between statistical g and a thing out there in the real world does not proscribe the possibility of there being a psychological theory of general intelligence. What is the

case is that this theory cannot rely on the mathematical abstraction g as its sole *raison d'être*. It is, therefore, crucial to find evidence to corroborate the reality of general intelligence, independently of the psychometric data base which gave rise to the concept.

Is there any evidence beyond the data matrix of intelligence test scores from which the notion of general intelligence derives that argues for its reality as a psychological construct? General intelligence cannot, by definition, be specific to any domain of knowledge. Thus it must be either a function of a cognitive control process that is involved in all domains or a non-cognitive physiological property of the brain. In either case it should be possible to find correlates of general intelligence in tasks that are relatively *knowledge-free*. The last ten years have seen a plethora of reported relationships. To the extent that they can be considered as valid indices of low-level information processing, they provide independent confirmation of the reality of general intelligence. The interpretation of these results is not without its problems, as we will see in the next chapter.

What can we Conclude from Psychometrics?

The main aim of this chapter has been to establish that:

1 We must maintain the clear distinction between intelligence tests as scientific instruments and the history and politics of intelligence testing. An abhorrence for this history should not cloud our judgement about *what such tests tell us about the nature of human intelligence.*

2 Since Binet, we are locked into the idea that *intelligence is a property of complex thought*. There has been no serious challenge to Binet's conception, at least from what might be labelled the 'anti-IQ' school. Ironically, the challenge that complex thought is fundamental to (rather than reflective of) intelligence has come from the major proponents of the scientific value of intelligence tests, particularly Arthur Jensen and Hans Eysenck. This is largely due to their view that intelligence is best conceived of as Spearman's g, or general intelligence.

3 Clearly *a theory of intelligence must address the issue of general intelligence*: that is, the fact that performance on a wide variety of tests is correlated. While it is true that the psychometric consensus is that there are group factors in addition to g, the latter is by far

the most important latent variable in the psychometric analysis of abilities. I hope the reader is convinced by now that, at the very least, there is a strong case for believing that psychometric g corresponds to some psychological attribute of thinking, rather than being, as some would have it, the artifactual creation of the technology of psychometrics.

The Limitations of the Psychometric Approach

There can be little doubt that intelligence testing provides an important data base for a theory of intelligence. Equally, there can be little doubt that the statistical armoury of psychometrics provides provocative taxonomies of abilities but cannot substitute for a psychological theory; the nature of psychometric intelligence is too volatile for that. For example, P.E. Vernon has emphasized the arbitrariness of the general intelligence/group factor distinction. The hierarchy of abilities (with g at the apex) does not stop at the group factors, as each of the group factors can itself be subdivided into at least two even more specific levels of ability. Vernon (1950) even ventured to suggest that the group factors may be 'almost infinitely subdivisible, depending only on the degree of detail to which the analysis is carried out'. Even more worrying is the realization that

> Indeed by including sufficiently similar tests, any specific factor (in Spearman's sense) can become a group factor. . . . It is even possible, when analyzing specialised tests, for a specific factor to become a general one. For example, a reaction time test analyzed in a battery of paper and pencil mental tests might obtain a g and a major group factor variance of about 10 per cent, specificity 90 per cent. On including two other kinds of reaction time, a small group factor would appear, while in a battery consisting only of such tests a general reaction time factor with 30 per cent or higher variance might be found; and we should be unaware that this was composed partly of 'higher' factors such as g and $k{:}m$. Thus there is no absolute distinction between general and specific factors as Spearman believed. (Vernon 1950, p. 26)

Clearly, then, a psychometric description is no substitute for a theory. Although many of the more scientifically orientated psychometricians recognize this (see, for example, the discussion in Jensen 1987a), all too often the main focus remains the search for the ultimate factor-analytic solution (Gustafson 1984). While it is easy to extract factors, give them names, and then pretend that they correspond to constructs in a theory (see, for example, the proliferation of such factors/constructs in Horn 1986), a theory of intelligence constitutes more than redescriptions of the data. In many of the psychometric debates about the nature of intelligence, no clear distinctions are made between 'ability', 'competence', 'skill', and so forth. This has led to many fruitless debates about how many abilities there are. No wonder. Without a theory of what an ability is, the question itself is pointless. There is no consideration of whether g, or general cognitive ability, is the *same kind of thing* as a specific ability, or, if they are different, what their relationship is. The only relationship that is discussed is that of relative importance in accounting for amount of variance in a test battery. General ability is usually regarded simply as some kind of addition to specific abilities (this is, in fact, an assumption of most factor-analytic techniques). This seems flatly contradicted by new evidence suggesting that g and non-g abilities have a differential relationship with changing IQ (Detterman and Daniel 1989). Even more depressing for the theorist is the fact that, although there are massive changes in intellectual competence through childhood and the whole practice of intelligence testing was based, originally, on this very phenomenon, there is virtually no developmental story to psychometric intelligence. We do not know, for example, whether the 'structure' of abilities changes during development, although most researchers assume that there is merely some kind of all-round improvement.[7]

This sad state of affairs is a necessary consequence of the essentially atheoretical development of psychometry. Psychometrists have concentrated on what they know best: measuring differences. We are well informed on sex differences, social class differences, race differences, and the effects of differences in schooling, nutrition, and just about anything else the imaginative psychometrist can think of. Intelligence is operationally defined as a score on an intelligence test. Factor analysis attempted to burrow beneath the test scores to generate theories of what lay underneath. But what lie underneath are factors, not psychological mechanisms. Factorial theories are theories about the structure

---

[7] To be fair, there was some psychometric work done in the 1930s and 1940s to look at the change in g loading of test batteries with age. This was not driven by any particular theoretical concerns, and represents a tiny fraction of the psychometric output on the structure of abilities. Since then there has been some impressive work on the influence of pubertal changes on spatial ability, although this has never been related to concepts of g, or indeed to development generally, most speculation being confined to an interest in sex differences.

of correlation matrices of ability tests; they are not theories of human psychology. Factors are abstractions from population variance, and cannot simply be given the status of psychological constructs without a *theory* relating a psychological mechanism to a factor. This applies just as much to Thurstone's PMA as to Spearman's *g*. A theory for the psychometrist has consisted simply in giving a factor a psychological-sounding name, like verbal ability, mental energy, or innate cognitive ability.

The other side of the coin is that there is nothing special about psychometric factors in this regard. Just because a construct such as general intelligence begins its life in psychometric measurement does not mean that it can have no psychological reality. The danger of the 'danger of reifying *g*' argument is that we make a logical error. To say that a factor does not *necessarily* correspond to a psychological attribute does not mean, correspondingly, that it cannot.

In the next chapter we will review the evidence from outside the domain of psychometrics for the psychological reality of general intelligence.

3

The Intelligent Synapse

In chapter 2 we examined the case for supposing that intelligence is not simply the invidious invention of politically motivated psychometricians. We concluded that the evidence in favour of the view that there is a large *general* influence on individual differences in intelligence is too robust to ignore. However, the case for general intelligence has been based, so far, solely on psychometric data. It is now time to consider evidence that not only argues for the psychological reality of psychometric *g*, but also attributes general intelligence to a *low-level* parameter of information processing. This will lead us inexorably into the debate between low-level and high-level views of intelligence introduced in the first chapter. We shall see not only that traditional cognitive views are antithetical to the notion of general intelligence, but that they are so because the cognitive framework inevitably regards intelligence as a *high-level* property of *knowledge systems*. The data we will deal with in this chapter are the correlations between performance on information-processing tasks and performance on intelligence tests. It will be useful to bear in mind a shorthand for the two competing explanations for the correlations. The *strategic hypothesis* has it that correlations between simple information-processing tasks and measured intelligence are based on knowledge-driven information-processing strategies that are more readily available to more intelligent subjects. The *speed of processing* hypothesis has it that correlations between simple information processing tasks and measured intelligence are based on their shared reliance on processing speed.

The strategic hypothesis corresponds to the view that intelligence is an experientially driven, high-level property of thinking and cognition; whereas the speed of processing hypothesis corresponds to the contrary view that general intelligence is a low-level attribute of information processing. Indeed, the most extreme form of the speed of processing hypothesis argues that this attribute of information processing is so low-level that it is parsimonious to think of it as a property of the physiology of the neocortex.

Intelligence as the Processing of Information

Cognitive Correlates

Cognitive psychology has long pursued the idea that the enormous variety of cognitive routines available to an individual can be decoded into an organized sequence of a small set of 'basic' information processing components; that, just as DNA is composed of just four chemical bases, so cognition is constructed out of combinations of a few basic processing components. This led to the obvious speculation that it was at the level of the efficient, or fast, operation of these components that the major differences in cognitive ability lay. The most concerted, and successful, effort in this tradition, collectively known as the *cognitive correlates* approach, is the work of E. Hunt and his colleagues in the 1970s (E. Hunt 1978, 1980; E. Hunt *et al.* 1975). The principal goal of their research was to find out which processes in Hunt's theory of verbal memory most differentiated 'high verbal' from 'low verbal' college students. The basic idea was straightforward. By the 1970s cognitive psychologists were confident that they had discovered a number of techniques that would allow them to measure basic operations in the information-processing system. For example, the experimental paradigm first developed by Posner and Mitchell (1967) was thought to provide an index of the time taken to retrieve information from semantic memory. Subjects make judgements about whether pairs of letters (for example, AA or Aa) were the same or different. They made the judgement under two different conditions, a name-identity condition (Aa = 'same') and a physical-identity condition (Aa = 'different'). The longer reaction time found for the name-identity condition was attributed to the extra time required to retrieve the name code from long-term memory. Cognitive correlates research then looked to see whether the efficiency of this hypothetical process correlated with verbal ability as measured by standard psychometric tests. Other paradigms were similarly used (for example, short-term memory scanning (S. Sternberg 1966) and sentence verification (Clark and Chase 1972) in an attempt to find out what operations best predicted psychometric verbal ability. Hunt himself considered that, as an attempt to isolate the major influences on individual differences in intelligence, this approach failed:

> While these studies should aid in advancing our understanding of the relationship between psychometric and information processing theories, the results to date do not indicate that they will produce a major breach in the 0.3 barrier [correlation between processes and psychometric scores]. They may push it back to 0.4, but the search for a 'true' single information processing function underlying intelligence is likely to be as successful as the search for the Holy Grail. (E. Hunt 1980, p. 457)

According to Hunt, the reason why variations in a single, basic information-processing component will never explain more than about 10 per cent of the variance in psychometric intelligence is that intelligence tests require the orchestration of several different processing components. In other words, the major individual differences are to be found not in the operating characteristics of processing components (the speed of processing hypothesis), but in how those components are selected and organized to solve a particular problem (the strategic hypothesis). Hunt's conclusion that intelligence is not a function of low-level properties of the processing system, but a property of processing *strategies* and *metacognitive* functions, set the scene for the work that was to follow in the 1980s.

Cognitive Components

The most influential and prolific of current theorists is Robert Sternberg. There have been two stages in the development of Sternberg's theory. The first stage involved developing a theory of the cognitive components involved in intelligent problem solving (R. J. Sternberg 1983). But even as early as 1984, Sternberg considered this 'cognitive view of intelligence' to be too limiting. Consequently, he further developed the componential theory to consider the cultural context of intelligence (contextual subtheory) and to specify what tasks required intelligent thought (two-facet subtheory). The result was his triarchic theory (R. J. Sternberg 1984, 1985). Because the earlier componential theory makes clear distinctions between high-level and low-level information-processing routines, it appears to provide an ideal framework for evaluating the basis of the major sources of individual differences. As a result, I will concentrate on Sternberg's conclusions about intelligence from the standpoint of his componential theory. The triarchic theory has, in any case, pre-empted the issue by shifting the burden of explanation of differences in intelligence away from possible low-level correlates of information processing towards the cultural shaping of intelligent behaviour.

Sternberg's componential theory derives from standard information processing *methodology*; in particular, the facility for analysing cognitive tasks into the hypothetical information-processing components necessary for their solution. Although E. Hunt had already hit on the idea of measuring different processes and assessing their contribution to individual differences in intelligence (by correlating their operating characteristics with standard psychometric measures), it was Sternberg who adopted the obvious research strategy (all good ideas seem obvious with hindsight) of analysing performance on the actual kinds of problems that comprise many standard tests of intelligence. Roughly, the basis of this research strategy is as follows. If intelligence tests measure intelligence and we want to understand what parts of the information-processing system contribute to differences in intelligence, what better way to find out than by constructing information-processing models of subjects' performance when solving intelligence test items? When we have an appropriate model, then we can see where in the model the major individual differences lie. An example may make this research strategy clear.

Analogical reasoning problems appear in many intelligence test batteries. For example, we might be asked to solve the following problem:

Lawyer   is to   client

as

Doctor   is to   ?

(a) medicine; (b) sick person.

Sternberg analysed the information-processing components involved in solving such problems by measuring the reaction time required to select the correct answer under different presentation conditions. For example, subjects may be pre-cued, before being presented with the full analogy, with the term *Lawyer*. If subjects who are pre-cued with this information are subsequently faster at picking *sick person* as the solution to the analogy, this may represent time saved in having to *encode* one of the terms in the analogy. Varying the pre-cued information and observing the effect on reaction time allowed Sternberg to model subjects' performance in terms of the operation of a number of information-processing components.

Sternberg's components are elementary processes invoked during problem solving. Each component has associated properties: how long it takes to complete its job (duration); how likely it is to return an error (difficulty); and the probability that it will be used. The components and their properties are derived by fitting information-processing models with different parameters to reaction time data. The componential theory posits that there are three kinds of information-processing components.

Performance components are those used in the execution of particular processing strategies. In the analogy example given above, they are components such as *encoding*, by which the subject recognizes the terms of the analogy, or *mapping*, by which the subject works out the higher-order relation between the two halves of the analogy. When the subject solves a problem, he or she is, in effect, invoking a sequence of performance components.

Metacomponents are responsible for selecting and monitoring a particular sequence of performance components that are appropriate for a given task. For example, there is a metacomponent responsible for 'recognition of just what the nature of the problem is' and another for 'understanding of internal and external feedback concerning the quality of task performance' (R. J. Sternberg 1983, p 5). In a nutshell, meta-components control processing strategies.

The third kind of component is a *knowledge-acquisition* component. Such components are responsible for learning new information and storing it in memory. The principle knowledge-acquisition components are (a) *selective encoding*, responsible for sifting out relevant from irrelevant information; (b) *selective combination*, responsible for organizing information that maximizes its internal coherence; and (c) *selective comparison*, responsible for relating the new information to that already stored in memory.

Sternberg regards these components as highly interactive. When faced with a problem, metacomponents select processing strategies and assemble the sequence of performance components that will instantiate this strategy. The relationship between metacomponents and performance components is hierarchical: metacomponents control performance components. However, the operation of metacomponents, and their success in selecting appropriate processing strategies, depends on the knowledge available to the system, which is influenced by the knowledge-acquisition components. They, in turn, depend on the problem-solving routines, selected by the metacomponents and implemented by the performance components, that are necessarily involved in acquiring much of our new knowledge. Although these processing components are interactive, Sternberg claims empirical evidence for their functional autonomy; in particular, how they relate to individual differences in psychometric intelligence. Thus, a cornerstone of information processing methodology, which reached its zenith with the additive factors method of S. Sternberg (1966, 1969) but dates back to the Donders' subtractive procedure in the last century, has been applied to understand how subjects solve intelligence test problems; but to what end?

The pertinent question is: What processing components most differentiate those who score well and those who score poorly on intelligence tests? It could be, for example, that the difference lies with the operating characteristics of the performance components. If so, then this would provide support for a low-level (for example, speed of processing) hypothesis. On the other hand, if the major differences are in metacomponent functioning, this would argue for a high-level, knowledge-driven, processing strategy hypothesis.

R. J. Sternberg, like E. Hunt before him, came to the conclusion that although performance components contribute to individual differences in intelligence, their contribution is rather weak overall. The major individual differences that seem to be most related to intelligence are those at the top of the hierarchy – the metacomponent processes – that is, the processes involved, principally, in organizing, rather than executing, problem solving. For Sternberg, then, intelligence should be considered, at the very least, as a high-level property of information processing. Where does this leave general intelligence?

To the extent that general intelligence resides in the components of information processing (and is not, for example, an exogenous property of cultural conventions), then, for Sternberg, it must be a property of metacomponent functioning. This must be so for two reasons: first, if general intelligence makes the single most important contribution to individual variation, it must map onto Sternberg's metacomponent functioning; second, metacomponents are the only processing components which exhibit any generality, since performance components, and to a lesser extent knowledge- acquisition components, are specific to particular problem types.

The cognitive correlates and cognitive components research has clearly favoured the 'high-level' conception of intelligence and argued against the 'low-level' one. However, there are a number of problems associated with both these research programmes.

(1)   Sternberg's componential theory of intelligence seems simply to be a redescription of the data. That is, it is not a theory of the intellect,

but a framework for describing problem-solving tasks. As Neisser (1983) has pointed out, the information-processing components given a particular label (for example, encoding) in one task may bear little relationship to the same component with the same label in another task. In other words, performance components give life to task descriptions, not mental operations. Such an approach inevitably leads to vacuous descriptions of what constitutes intelligence. For example, intelligence may be the property of a metacomponent which allows 'recognition of just what the nature of the problem is' (R. J. Sternberg 1983, p.5). However, this sounds like a simple restatement of the phenomenon we seek to explain. Again, as Neisser has put it:

> Things like 'Recognition of just what the nature of the problem is' and 'Understanding of internal and external feedback concerning the quality of task performance' [both metacomponent functions] are not separate elements in any genuine mental process; they are more like chapter headings in books on how to think. (Neisser 1983, p.196)

(2)   The early failure to find low-level correlates of intelligence is, if not an inevitable consequence of cognitive metatheory, then at least a highly likely one. I say so for the following reasons:

(a) The enterprise of examining problem-solving ability as a sequence or a hierarchy of information-processing components was always likely to be antithetical to the major psychometric construct of general intelligence, or g. This is because an information-processing description of problem solving is, by and large, task-specific. That is, a flow chart itemizing the kinds of components used and their organization and sequence of operation would differ radically from one task to the next. This means that it could only be the components at the highest levels in the hierarchy that would show the generality across tasks that would account for g (R. J. Sternberg 1983). Indeed, precisely because these descriptions do vary from task to task, this predisposed cognitive psychologists to disbelieve in g. So it should come as no surprise that if our metatheory requires us to analyse problem-solving tasks into separate stages of processing, we may be misled into believing that intelligence is composed of a number of independent information-processing mechanisms. Sternberg's framework may come to be regarded as a notorious example of confusing an information-processing flow diagram with an information-processing theory. The componential method has been very successful in describing how individuals may solve different kinds of intelligence test items, but this may be a case of not seeing the wood for the trees. Intelligence is not the aggregate of task performances; consequently, it is not the aggregate of all the information processing operations that subserve these tasks. Yet, for traditional cognitive psychology, it could hardly be anything else.[1]

(b) That the specificity of intelligence was assumed, rather than tested, by cognitive psychologists can be seen in the choice of relationships that were sought. Attempts were made to relate parameters of verbal processing to *verbal* intelligence, spatial processing to *spatial* intelligence, and so on. Only in the field of mental retardation was the notion of general differences in intelligence given even minimal credence. Similarly, if researchers had been really serious about looking for the sources of intelligence they would not have confined their subject samples mainly to college students. Students may be convenient subjects, but they are hardly a representative sample of our variable of interest.

(c) In any case, the possibility of finding relationships between low-level basic processes and intelligence depends on an appropriate description of the 'basic' processes. This is, of course, dependent on particular processing theories, theories which may be wrong. Thus, if there is really no such thing as a buffer X, then it is unsurprising that measures of read-out rates from buffer X (or whatever) do not relate particularly highly to psychometric intelligence. And it would be a brave cognitive psychologist who would claim to have worked out, even in the broadest terms, an uncontentious processing theory of any significant cognitive ability.

(3) It is not surprising that the standards of psychometric practice are usually ignored when measuring cognitive processes. These experimental measures were designed with a different job in mind, that of elucidating the nature of mental mechanisms. They were never intended to be 'tests' of those mechanisms. Given the likelihood of their diminished reliability compared with a standard psychometric test, and given the

---

[1] That we may have a framework dispute is highlighted by the admirable study of problem solving in Raven's progressive matrices by Carpenter *et al.* (1990). For many psychometrists, this test is taken to be a pure measure of *g*, the unitary trait underlying general intelligence. Yet Carpenter et al., unsurprisingly, show that the processes involved in solving these problems are many and varied. Whether this observation, by itself, could be used to invalidate the unitary conception of intelligence is clearly a separate issue. Its relevance is attenuated, for example, by the lack of correspondence between the processes identified as central to solving Raven's problems and those presumably involved in other high *g*-loaded tests such as tests of vocabulary. What these kinds of cognitive analyses do seem to show is that the generality of intelligence cannot be sought in the overlap between models of different task performance

restricted range of intelligence in the subjects used, it is no wonder that the correlations failed to crawl above 0.3.

In sum, many of the attempts to look for low-level correlates of intelligence had a half-hearted look about them – more a case of a methodology in search of a problem than a problem in search of a solution.

Reaction Time and Intelligence

Given the cognitive metatheory, it is no surprise to discover that those who now claim to have found E. Hunt's Holy Grail are not cognitive psychologists. Instead, they follow a tradition, inspired by Francis Galton in the last century, which sought to locate the primary source of intelligence in the speed of simple sensory processes. Although Galton was unsuccessful, his attempt gave rise to a conception of 'mental speed', which pre-dates the concept of information processing and whose very vagueness guarantees its longevity. In this tradition, there is, *a priori*, a lack of concern to specify the mechanism whose speed it is we are measuring. The 'mind' will do. As a result, this movement has shown an impatience with the pedantic concerns of the cognitive psychologist, who seems to be asking perennial questions about specific cognitive processes of dubious reality. Perhaps this impatience is born of the belief that the basic 'cognitive' process has been found in the processing of a 'bit' of information.

All the nineteenth-century work on reaction time (RT) and intelligence used what would now be called a 'simple' reaction-time paradigm, with the subject being required to respond as quickly as possible on detecting a stimulus. There is only one stimulus and only one response. Choice RT, on the other hand, involves selecting one of a number of possible responses, depending on the nature of the stimulus. Hick (1952) and Hyman (1953) noted that RT increases linearly with the $\log_2$ of the number of response alternatives. This is a linear increase with the number of 'bits' of information (Shannon and Weaver 1949). Eysenck (1967) cites a study by Roth (1964), who found that the slope of this linear function correlated with IQ; the steeper the slope (indicating the increase in time to process more information), the lower was the subject's IQ. Eysenck (1967) used this evidence as a central plank in his theory claiming that 'mental speed', now thought of as the rate of processing bits of information, underlies differences in g.

While Eysenck was the first to resurrect Galton's pursuit of mental speed and intelligence, it was Jensen who developed both the empirical and the theoretical basis of this association (Jensen 1980, 1982, 1987b). His methodology has become the standard, and to understand much of the controversy surrounding the relationship between RT and intelligence, we must consider it in some detail.

Figure 3.1 *Jensen's Reaction Time task. The subject lifts his finger off the home button to press the button below the illuminated target.*

In Jensen's procedure, subjects rest the index finger of their preferred hand on a 'home' push button (see figure 3.1). Arranged in a semicircle around the home button are eight push-buttons, each with a light above it. When one of the lights is illuminated, the subject must move as quickly as possible from the home button to the push-button immediately below the light. RT is taken as the time from the onset of the stimulus light to the time when the finger is lifted from the home push-button. Movement time (MT) is the time taken to push the button below the stimulus light after stimulus onset, minus RT. In this way RT and MT are experimentally independent. The major independent variables in these studies are IQ and the number of 'bits' of information in the task (the number of stimulus lights; two, four, and eight lights representing one, two, and three bits respectively).

The major results claimed from numerous studies are as follows:

(a) intra-individual variability in RT (rather than mean reaction time) is the single best predictor of IQ, lower-IQ subjects being more variable (Jensen 1980, 1982); (b) the difference between IQ groups widens as the number of bits increases (Jensen and Munro 1979, Jenkinson 1983, Lally and Nettelbeck 1977, Smith and Stanley 1983); (c) RT parameters discriminate between IQ criterion groups – for example, groups with retarded and normal IQ (Nettelbeck and Kirby 1983, Jensen 1982, P. A. Vernon 1981). In short, since Jensen considers his task to have a minimal knowledge content, he attributes the RT/IQ relationship to their shared reliance on processing speed. High-IQ subjects process each bit of information faster than low-IQ subjects.

Jensen has proposed a theory, the details of which need not concern us here, which relates processing speed, and therefore g, to the phase of excitatory-refractory oscillations of neurons. Jensen's conclusions about intelligence clearly contradict the views of R. J. Sternberg and E. Hunt outlined at the beginning of the chapter. The correlation between performance on information-processing tasks and psychometric measures of intelligence is based on differences in speed of processing rather than their shared reliance on processing strategies. Not only is there a single process that underlies differences in intelligence, but this process is a low-level property of neuronal physiology, rather than a high-level property of cognition.

Criticisms of the Speed of Processing Hypothesis

The idea that differences in intelligence do not depend on complex cognitive processes is a serious challenge to a cognitive view of intelligence. It is no surprise, therefore, to discover that Jensen's views have not gone uncontested. Understanding the debate between Jensen and his critics requires a more detailed consideration of his methodology. It is worth spending some time on this, because the terms of reference of this debate cut to the heart of the cognitive approach to understanding intelligence.

Jensen's theory is based on strong assumptions inherent in his RT paradigm. The major assumption is that the RT task indexes only processing speed and is not influenced by differences in knowledge or knowledge use. This conclusion has been vigorously challenged (Longstreth 1984, 1986; Rabbitt 1985; Detterman 1987). The challenges centre on the contrary assertion that performing well in a reaction-time task is actually a very complicated business. Rather than RT reflecting the speed of operation of something as straightforward as the processing of a 'bit' of information, it reflects the operation of a complex processing system which, itself, has many different facets, including attention, motivation, visual search, and encoding. Differences in RT, Jensen's critics suggest, may reflect differences in one or more of these diverse processes or, even more congenially for the cognitivist, in the organization and control of these processes.

Longstreth claimed that Jensen's studies abounded with procedural flaws which confounded many variables. For example, the effect of complexity (number of bits) was often confounded with order of trials, introducing possible practice effects; complexity was also confounded with visual attention variables, with more complex arrays requiring wider scanning patterns (see Jensen and Vernon 1986 and Longstreth 1984). Therefore, it may be not that low-IQ individuals are *slower* than high-IQ individuals, but that they do not benefit as much from practice or have poorer visual scanning strategies. Longstreth argues, then, that differences in RT could be caused by differences in *high-level* cognitive structures underlying differences in processing strategies.

Further, Rabbitt (1985) claims that the reaction-time task only *appears* to be simple, and that hidden beneath the surface is a complex information-processing problem. The common instruction to 'go as fast as you can but make as few errors as possible' sets up a complex control problem in which speed and accuracy must be traded off to obtain optimal levels of performance. This involves monitoring and controlling speed of response in relation to varying probabilities of errors. As a consequence, increased mean RT and, particularly, increased variability in RT associated with lower intelligence are attributable to less efficient organization of cognitive routines rather than decreased efficiency, or speed, in the operation of processing components. Parameters of RT correlate with IQ because they have something in common: not the influence of variations in neuronal speed, but the quality of organization of knowledge systems.

Certainly, RT data abound with the hallmark of strategic influences. For example, Nettelbeck and Kirby (1983) claim to have evidence that the retarded may, contrary to instruction, lift their fingers off the home button before selecting which light to move to. Therefore, differences between these groups may be a reflection of strategic differences, rather than differences in processing speed. The fact that the most conflicting results in the Jensen paradigm concern the effect of complexity is consistent with the idea that changes in complexity would be likely to increase the opportunity for variations in task strategies.[2] Lally and

---

[2] See Jensen (1982) for a comprehensive review of the literature. In that paper Jensen admits that the data on the relationship between complexity and RT variability are consistent with his neutral oscillation model.

Nettelbeck (1977) and Smith and Stanley (1983) found, as would be predicted by Jensen, that the strength of the correlation between IQ and RT parameters was greatest at an information load of three bits (eight choices). On the other hand, Nettelbeck and Kirby (1983), reanalysing the Lally and Nettelbeck (1977) data, conclude that within homogeneous IQ groups there is no such increase with complexity, a result substantiated by P. A. Vernon (1983) and Jenkinson (1983). Jensen himself (1982) acknowledges that the linear relationship between RT and complexity breaks down for RTs over 1 second at an information load of about four to five bits. He attributes this to the invocation of different mental processes to cope with the increased information load on the brain's limited information-processing capacity. However, cognitivists would claim that these processes are invoked at all levels in RT tasks, and that it is the variance in their use – in other words, variance in cognitive processing strategies – that creates the correlation with IQ, and not the variance in processing speed. Thus the major assumption of Jensen is called into question. The correlation between RT and IQ is created by the intelligent use of cognitive processing both in RT tasks and in intelligence tests.

What can we Conclude about Reaction Time and Intelligence?

We are probably in danger of becoming overly concerned with the minutiae of experimental detail, and this may be the time to stop and consider the central issue. Remember that the importance of the RT/IQ association was that it was thought to support the hypothesis that the basis of individual differences in intelligence lies in differences in speed of processing. This was based on the assumption that the RT task is knowledge-free – that is, that it is so simple that variation in performance cannot be due to differences in knowledge. The question is: How seriously should we treat the counter-argument that the standard reaction-time task is actually very complex, and that it is the co-variation in task complexity that generates the RT/IQ test correlation?

Conversely, the claim that the RT task is rather complex might be treated with some scepticism, given that the basis of the original opposition from cognitive psychology to the idea of an RT/IQ correlation was that an RT task is too simple to be related to intelligence. Yet, there can be no doubt that the RT task is very much more complex than Jensen originally supposed. However, if the claim is that Jensen's RT procedure actually involves the same level of complexity of operations as a typical intelligence test item, then the claim is clearly preposterous.

For example, it is not the case that the mentally retarded are incapable of producing accurate responses in the RT paradigm, as they clearly are for many examples of intelligence test item; it is just that they take 100 milliseconds longer, on average, to do so. Further, it is hard to imagine anyone arguing that the ability to trade off speed and accuracy in an RT task directly contributes to the process of acquiring a large vocabulary, or knowing who wrote *Faust*, or solving a difficult matrices item, or any other of the myriad forms that intelligence test items take. What the RT data demonstrate is not that intelligence is a collection of arbitrary pieces of knowledge and cognitive control processes, with 'RT ability' as a heretofore undiscovered one, but that, at the very least, intelligence must be a processing parameter underlying knowledge, knowledge use, and knowledge acquisition. This view is reinforced by an even simpler task which, it is claimed, shows even stronger relationships with measured intelligence.

Inspection Time and Intelligence

The straightforward idea behind the attempt to relate RT to intelligence is the belief (which turns out to be false) that the RT task is so simple that there is only one way of doing it. If there are no strategic variables in an RT task and yet it relates to differences in knowledge-rich, intelligence test performance, this must mean that intelligence is a more basic aspect of processing than is knowledge itself. In particular, the relationship between RT and IQ lends weight to the idea that intelligence is based on speed of processing. However, as we have seen, RT shows some classic features of strategic influences. This does not mean, of course, that the correlation between RT and IQ is not based on a shared relationship with processing speed. But it no longer makes it the only explanation. However, the discovery of another measure, where even the modest strategic influences present in an RT measure are absent, reinforces the speed of processing hypothesis.

Inspection time (IT) is hypothesized to reflect the minimum time required to make a single inspection of the sensory input (Vickers and Smith 1986). The task used to measure IT is even simpler than that used to measure RT, because there is no requirement for fast responding, only for accuracy. Speed, in this task, is an attribute of the stimulus, not a dependent variable. In the standard IT paradigm a subject is presented with a stimulus consisting of two vertical lines, of markedly different lengths, joined at the top by a horizontal bar. The subject must simply indicate whether the longer line is on the left or the right

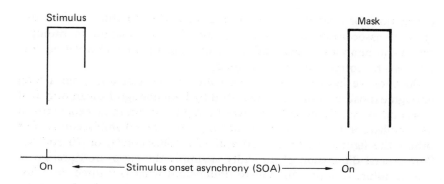

Figure 3.2 *A stimulus is presented for a variable interval before the onset of a masking stimulus which prevents further processing of the stimulus. The stimulus can have the long line either on the left or on the right; here it is on the left.*

of the stimulus (see figure 3.2). The exposure duration of the stimulus is varied by changing the time of onset of a masking stimulus, which obscures the original stimulus and prevents further information from being extracted. Shorter exposures result in more errors. At very fast exposures, performance drops to chance levels. IT is an estimate of the shortest exposure at which performance is virtually error-free. This estimate varies among individuals and is negatively correlated with IQ. That is, more intelligent subjects can maintain error-free performance at shorter-exposure durations than less intelligent subjects.

Nettelbeck and Lally (1976) found a correlation of -0.92 (p < 0.01, two-tailed) between IT and Performance IQ based on the Wechsler Adult Intelligence Scale (WAIS), but only a non-significant correlation of -0.41 with Verbal IQ. Since those heady days, the correlations have been less dramatic. In Adelaide, Nettelbeck and his colleagues have reported a range of studies in which the (negative) correlations between IT and IQ have ranged from the low 0.2s to the high 0.9s (Nettelbeck *et al.* 1984, Nettelbeck 1982, Lally and Nettelbeck 1980). Brand (1980) and Brand and Deary (1982) summarize a collection of small studies conducted in Edinburgh which have reported (negative) IT/IQ correlations ranging from the low 0.4s to the high 0.9s. Although the correlations are variable, the Adelaide and Edinburgh studies invariably find a negative and usually significant relationship between IT (which has been measured in different ways using different apparatus and sometimes involving different sensory modalities) and some form of IQ measure. Nettelbeck (1987), summarizing the data from 29 studies, concluded that a realistic estimate of the size of the true population correlation between IT and IQ is around 0.5. This is a substantial correlation for a single processing measure. RT variables achieve this level of association only by being used in combination in multiple regression analyses (leaving aside the dubious practice of inflating correlations by correcting for attenuation).

So here we have a task that virtually anyone can do, given a long enough exposure to the stimulus. And by long enough I mean only half a second or so. Some subjects will be highly accurate at exposures of 100 milliseconds but will start making errors at 80 milliseconds. For others the figures might be 130 and 120 milliseconds, or 70 and 55. The crucial dimension of difficulty is simply the exposure duration of the stimulus. It is not difficult to understand what is wanted, it is just that sometimes the stimulus appears too fast to be 'seen'. There is no need to trade off speed and accuracy, since subjects can take as long as they like to report their decision. The speed variable in this case is, as mentioned above, not within the control of the subject, but is an attribute of the stimulus. Is this, then, the task that has shown, irrefutably, that intelligence is based on differences in speed of processing? Drearily, yet predictably, the answer is still no.

Criticisms of the IT Studies

There are more complications with the concept and measurement of IT than was at first thought. These complications are worth considering in some detail, for two reasons: first, because they again highlight the nature of the dispute between those who believe that differences in intelligence could be based, at least in part, on differences in low-level mechanisms and those who believe that differences in intelligence are a consequence only of differences in high-level cognitive systems; and second, because IT will form a central part of the data base from which empirical support for the theory advanced in this book will be sought, which means that more detail on its strengths and weaknesses as an index of low-level processing speed, or efficiency is needed.

There are two principal criticisms of the IT work: first, that the studies used too few subjects and that too many subjects were mentally retarded; second – and this is the same as the argument used against the RT work described above – that the task in which IT is measured is much more complicated than at first appears.

Certainly the early studies used small samples, with non-normal distributions of IQ, and usually included mentally retarded subjects, factors likely to result in overestimation of the size of the true correlation with intelligence for the population overall (see Mackintosh 1986). Indeed, it has been argued that the correlation depends solely on the inclusion of mentally retarded subjects. In effect, the claim is that there is no correlation, just a difference between normal and mentally retarded subjects, and normal subjects perform better than mentally retarded subjects on just about any task. It is true that some studies which excluded subjects of subnormal intelligence have found low or insignificant correlations (Hulme and Turnbull 1983, Irwin 1984). However, there have been many studies since then which have used representative samples of subjects and have found significant correlations between IT and intelligence. For example, Nettelbeck (1985) reported correlations of −0.38, −0.55, and −0.46 with Verbal, Performance, and Full Scale IQs from the WAIS, using 40 adult subjects in the normal range of IQ. Similarly, Anderson (1986a) reported a correlation of −0.41 with WAIS Full-Scale IQ for 43 school children attending a normal school, and a correlation of −0.39 with Raven's Matrices for 113 normal schoolchildren (1988). Finally, Longstreth, who had been one of the most dogged critics of the RT work and who was a self-proclaimed sceptic of the IT research, reported with his colleagues in 1986 a correlation of −0.44 between IT and IQ with a sample size of 81, and admitted 'We are "Doubting Thomases" no more.' Recent 'meta-analyses' of a large number of IT studies, taking into account issues of sample sizes and range of IQs, estimate the 'true' correlation between IT and IQ at about 0.5 (Nettelbeck 1987, Kranzler and Jensen 1989). It is fair to say that the criticism that the IT/IQ relationship is based purely on the inclusion of mentally retarded subjects, or rectangular (rather than normal) distributions of IQ in the subjects used, is no longer tenable.

What, then, of the argument that IT, like RT, is more complicated than was first thought? Undoubtedly, the model of sensory discrimination underlying the first conception of IT by Vickers (1970) has been complicated substantially by more recent data. These complications have cast doubt on the notion that IT is an index influenced only by the speed of early sensory processing. The early model assumed that a single inspection of a typical IT-type stimulus must favour the correct response as long as the discrimination to be made involved a difference in visual angle of 0.8°, a stimulus whose sensory effects are far greater than the noise level in the system. However, a study by Vickers and Smith (1986) suggested that changes may occur in sensory discrimination, due to adaptation to the cumulative distribution of sensory magnitudes presented over trials, which could result in even an easily discriminated stimulus generating an ambiguous inspection. In such a case a further inspection of the stimulus would be necessary. If true, this would mean that the standard IT task may overestimate the value of IT because the estimate may be based on more than one inspection. It may be, then, that individual differences in a variable such as adaptation contribute to IT, as well as processing speed. However, it must be said that any such effect has still to be shown for IT-type stimuli. At the moment this is just a theoretical possibility. Further, any effects of adaptation must be marginal, and are unlikely to contribute significantly, for example, to the large differences in IT reported between normal and mentally retarded subjects. Nevertheless, IT must now be considered to reflect more than just speed of processing; it probably has a sensory adaptation component as well.

A related complication is the possibility that different subjects use different criteria in responding. Reaction time is often measured 'incidentally' in an IT study. For normal subjects, even though told that accuracy is more important than speed, their RT will increase as the stimulus exposure duration decreases, until an asymptotic value is reached at the level of an individual's estimated IT. For retarded subjects, on the other hand, RT remains fairly constant across all exposure durations. This may be indicative of *deadline responding*. That is to say, the mentally retarded subjects will respond after a certain amount of time, irrespective of the state of the evidence. Normal subjects, on the other hand, respond at a rate that matches the accumulation of evidence for one stimulus or the other. Despite the possibility of these different 'strategies' no one has yet demonstrated their consequences for the estimate of IT in the two groups, or whether such differences exist in the normal groups of subjects tested.

Time out

Again, we are delving (necessarily) into the minutiae of experimental detail. Before we draw a general conclusion about the status of the IT research, I want to point to a general trend running through the criticisms of the low-level correlates of intelligence research. As the putative task that indexes the speed of 'basic' processes becomes more and more 'simple', starting from processing components in particular information-processing models through speed of processing of 'bits' of information (RT) to speed of early sensory encoding (IT), so do the complications and criticisms become more detailed and low-level. Remember that what is interesting about these tasks is the reported relationships with psychometric intelligence and the contrast between high-level explanations (intelligence is a function of knowledge, metacognitions, and so forth) and low-level explanations. Although these progressively simpler tasks are not as simple as they at first appear, the ways in which they become complicated simply put more emphasis on variation in low-level parameters of processing. We cannot be sure that we have isolated *the* low-level parameter that correlates with intelligence, but we can now be sure that there is, at least, one. For example, it is highly implausible to attribute sensory adaptation and response criteria effects in an IT task to anything other than the fine tuning of extremely detailed and low-level sensory mechanisms. This takes the sting out of many of the criticisms centred on the idea that these tasks, rather than being simple, are actually rather complicated. They are complicated only in the sense of involving complicated mechanisms. The only sense of 'complicated' that is critical for the essence of the high-level argument is if knowledge regarding the nature of 'simple' tasks can be used to find *other ways* (different strategies) of doing them. This would open the door to intelligent people using *better* strategies, thereby artifactually producing a correlation between a so-called low-level index of processing and intelligence. If, on the other hand, a simple mechanism has more parameters than we first thought, and, crucially, those parameters are not within the strategic control of the subject, then this does not invalidate the proposition that low-level mechanisms vary in speed or efficiency with level of intelligence. Is this the case for inspection time?

Back to inspection time

Despite the new complications to the original simple model in which IT reflected speed and nothing else, we find that IT is very much a *psychophysical* task. That is, it is a task that has well-defined psychophysical functions and whose main parameters are influenced by sensory/perceptual variables, rather than by cognitive ones. For example, despite possible differences in criteria for responding, individual psychometric curves relating exposure duration to accuracy conform to a cumulative normal ogive. This function is not only found for many other psychophysical relationships; it was also predicted by Vickers's original theory. So the crucial dimension for IT is exposure time, and the relationship between accuracy and exposure time follows a precisely defined psychophysical function for all subjects (Nettelbeck 1987). Similarly, absolute levels of IT are influenced by psychophysical variables such as the relative brightness of the stimulus and the mask and the quality of the mask. It is hard to explain these kinds of effects with reference to anything other than a detailed sensory encoding model. What kind of knowledge variables could be brought to bear on the task whose influence would depend crucially on such factors as the brightness of the screen?

So IT is probably the processing measure least influenced by differences in strategic processes that we have yet encountered. The major differences in IT are probably caused by variations in the speed and efficiency of low-level processing mechanisms. Indeed, where strategy effects have been reported (MacKenzie and Bingham 1985, Egan 1986) they *detract* from the relationship with IQ rather than enhance it. For the moment, I simply want to emphasize that relative to RT, IT seems a better candidate as a measure of 'basic' processing speed. That it is not totally free from the influences of task strategies will become an important part of my argument when we return to IT in the context of developmental change. But, while the details of the RT and IT experimental procedures were being scoured for possible artifacts by those who wanted to reject any relationship between low-level processes and intelligence, the whole dispute seemed to be undercut by the apparent discovery that intelligence was to be found in the efficient operation of neurons. This is as low-level as we can go.

Evoked Potentials and Intelligence

The logic of those who seek low-level correlates of intelligence is to go to such a low level of processing that there is no possibility of influence from higher cognitive functions. Some regard the lowest level as parameters of neuronal functioning. If variations in neuronal functioning could be shown to be related to differences in intelligence, then this would demonstrate, once and for all, that intelligence is built upon differences in low-level properties of the nervous system and that differences in higher-level cognitive functions are epiphenomenal. This view is, of course, a declaration of faith, rather than the consequence of logic. The faith is based on a particular philosophy of science: *reductionism*. We will have to deal with this philosophy to understand what the data linking evoked potentials to intelligence may mean. But first the data.

The activity of the brain creates electrical fields that can be monitored by placing recording electrodes on the scalp. The changing electrical potentials are used to deflect a pen recorder to leave a trace of the brain's activity, known as an EEG (electroencephalogram) recording. The EEG can be monitored in a waking, normal subject who can be asked to perform a variety of tasks. The effects of those tasks will be recorded, simultaneously, in the EEG. The fact that cognitive processes (that is, tasks which require 'thinking') can have differential effects on EEG activity is well established in adults (Courchesne *et al*. 1975,

Donchin and Coles 1988, Martindale *et al.* 1984) and has been established more recently in infants (Hofmann *et al.* 1981). It is, of course, not surprising that different kinds of thinking should produce corresponding differences in brain activity that can be monitored by EEGs (though more surprising and less straightforward than one might at first imagine), since we all agree that thinking depends on brain activity. But the critical question is: Can we find parameters of the EEG wave-form that will distinguish *intelligent* from *unintelligent* thinking? Could it be that a simple parameter of EEGs can predict intelligence? The most obvious parameter of neuronal functioning that might be related to intelligence is speed of conductivity. Therefore, early attempts to find parameters of the EEG that correlated with IQ concentrated on the *latency* of evoked, or event-related, potentials.

An evoked potential is one that is produced in response to a stimulus (or, in some situations, in the event of an absent, but expected, stimulus). Often this stimulus is simply the flashing of a light or an audible tone in what would otherwise be a muted sensory environment. The EEG wave-form produced by this event can be analysed into components. They are usually identified with respect to their polarity (positive (P) and negative (N)) and their temporal relationship to the evoking stimulus (measured in milliseconds; for example N150, P300).

In the attempt to locate a speed of processing basis of intelligence, it seems obvious to look for relationships between individual variations in the latencies of these components and IQ. Correlations of around –0.3 have been reported (Ertl 1966, Ertl and Schafer 1969, Schucard and Horn 1972). The size of these correlations was discouraging, being no more than those found for many low-level cognitive measures. However, dramatic new relationships were discovered when the focus of attention moved from latency to form. The average evoked potential (AEP) is taken from several recordings made in response to the same stimulus. Subjects are given repeated trials, consisting, say, of a flash of light, and the wave-forms produced by this event are added to each other, producing an averaged wave-form. The Hendricksons and colleagues (Hendrickson 1982, Hendrickson and Hendrickson 1980, Blinkhorn and Hendrickson 1982) measured the AEP by placing a piece of string around the contours of the wave-form and subsequently measuring its length. The claim is that the length of this piece of string correlates with IQ at around 0.8 (see figure 3.3). These results have since been replicated by others (Haier *et al.* 1983, Robinson *et al.* 1984). They are, indeed, dramatic, and have left most of the strategy arguments levelled against the RT and IT procedures high and dry. But what do these results really mean?

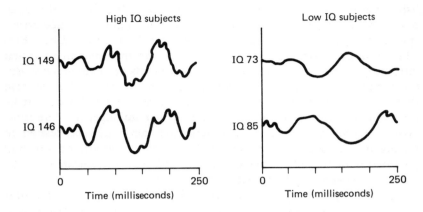

Figure 3.3 *Averaged evoked potentials (AEP) for high IQ and low IQ subjects. The high IQ subjects have 'spikier' waveforms (after Eysenck 1986).*

The rationale behind the string-length measure is that when any wave-form is added to any other, the peaks and troughs must occur in phase, otherwise they will tend to cancel out. The 'noisier' a wave-form is, the less likely it will be that the peaks and troughs generated by the event will be in phase over repeated testing. As a consequence, the AEP will tend to flatten out. For example, a series of recordings of the background level of neuronal activity (which is assumed to reflect only noise in the system) will produce a wave-form that will approximate a straight line when averaged. By contrast, a low-noise wave-form, faithfully reproducing the information content of a stimulus event, will produce a very 'spiky' wave-form when averaged (because peaks will be added to peaks, and troughs will be added to troughs). Therefore, the spikier the averaged wave-form produced in response to a stimulus (hence the longer the string length), the greater the fidelity in information transmission in the nervous system. Thus, because spikiness (string length) correlates with IQ, high IQ is hypothesized to be the result of superior fidelity in neural transmission. This concept is different from speed, but it is related.

Eysenck (1986) has proposed that speed is a second-order derivative of neuronal *efficiency*, as measured by the spikiness of the AEP.

Interestingly, Eysenck explains this with particular reference to RT data.[3] He assumes that because of the relatively low signal-to-noise ratio in the nervous system, information must be transmitted repeatedly. Some kind of device will be responsible for comparing successive transmissions and making the decision as to what the 'true' signal is. Given this, the less efficient the transmission process is, the longer, on average, it will take to determine what the signal is. It is easy to see that the time taken to make decisions (RTs) will also be more variable, but that occasionally even a low-efficiency system will allow a decision to be made on the basis of a low number of transmissions. This would make sense of the finding that, although longer mean RTs are associated with low IQ, low-IQ subjects occasionally produce very fast RTs (Jensen 1987a,b).

For our purpose, it does not matter that the assumptions made in this model are massive and, almost certainly, dubious. Nor would I wish to imply that the details are at all important for Eysenck's central proposal that synaptic efficiency is the basis of differences in intelligence. The point, for Eysenck, is to establish that, in principle, differences in neuronal efficiency could be used to explain detailed behavioural correlates of intelligence and that, therefore, it is the synaptic theory which will ultimately be the most appropriate for understanding human intelligence. By contrast, although I do not want to deny the possibility of a biological basis for general intelligence, I would deny that a theory of synaptic processes is the most appropriate level of explanation of human intelligence. But before I attempt to convince you of this, I must complete the descriptive cycle that has emerged in this chapter and present criticisms of what might otherwise be regarded as unassailable data in favour of the low-level view of intelligence.

Criticisms of the Evoked Potentials Work

It would be easy to be seduced by the simplicity of the evoked-potential paradigm. What could be simpler than asking a subject to sit quietly in a dark room and then recording what happens to his or her brain when a light is flashed? What is found is that those with complicated brainwaves turn out to be more intelligent. But all we did was flash a light; no thinking or problem solving or metacognizing here. So, the argument goes, fundamentally, intelligence cannot be anything to do with thinking. It must, therefore, be a product of physiology. However, things are not that simple.

[3] I say 'interestingly' because, as we shall see below, Eysenck still has to invoke a cognitive model to make sense of his hypothesis, and this seems to conflict with his 'cognition is epiphenomenal' stance.

Much of the force of this argument rests on the assumption that the evoked potential is a simpler, more basic, measure of processing than either RT or IT in the *real* nervous system. However, this is a huge assumption. Discovering exactly what the relationship is between the real nervous system – that is, neurons and their collective behaviour – and evoked potentials is an active research area in its own right. Evoked potentials are enormously complex epiphenomena of all sorts of features, and cannot be considered as direct indicators of cortical activity. Convolutions of the cortex, the interaction of electrical fields from multiple-generator sources, distortion of electrical potentials due to skull openings and varying thickness of the skull, and the influence of the conductive properties of the cerebral spinal fluid, meninges, dermis, and fat layers all contribute to variation in evoked potentials.[4] Given these many sources of variation, it is difficult to unambiguously attribute individual differences in evoked potentials to individual differences in information processing.

As far as providing data on cognitive processing is concerned, the interpretation of evoked potentials is probably more problematic than the interpretation of more behavioural indices such as RT and IT. They are certainly not simple measures of neuronal efficiency. Given this complexity, it is not obvious how an IQ/evoked potential correlation should be interpreted. The apparent simplicity of the paradigm makes it seem as if there are no mediating factors between the nervous system and intelligence. However, in reality, the AEP is just one more indirect index of processing (like RT and IT); so we should not be surprised to find that, as with the others, alternative explanations for the correlations have been put forward.

Thus, as with RT and IT, the suggestion is that the correlation between an evoked potential (supposedly a knowledge-free measure) and intelligence test performance may be attributed to variations in processing strategies, variations which may, in turn, be knowledge-based. Components of an evoked potential are influenced by such factors as the probability of occurrence of the stimulus and the attentional state of the subject. Indeed, Mackintosh (1986) has suggested that high-IQ subjects' stability in evoked potential over many trials may be due to the willingness of these subjects to comply with the instruction to do nothing, but nevertheless stay alert, during testing. So it is not the case that there is no room for strategic differences between the subjects to influence the data. However, it must be emphasized that it is often

---

[4] I would like to thank Mark Manning for helping to furnish this list of contributory factors.

the early components, such as the N140 and P200 (Haier *et al.* 1983), which are particularly related to intelligence in these experiments, whereas it is later components like P300 that seem to be more sensitive to 'psychological' variables. Further, the sheer size of the correlations is remarkable, although here too there may be some suggestion of artifact in the scaling of the voltage parameters (Vetterli and Furedy 1985). Although these data are provocative, they have a central weakness. Unless we understand the relationship between EEG components and neurons on the one hand and EEG components and cognitive processes on the other, the correlations between EEG components and IQ scores do not help us to understand the relationship between neurons and cognitive processes.

But for the moment let us take these data at face value and suppose that intelligence is based on *some kind* of property of information transmission in the nervous system. Is it our job, as theorists of intelligence, to find out what its physiological basis might be? We now meet the major objection to these data. Their usefulness depends on a belief in the possibility of reducing psychological constructs to physiology.

Reductionism

Eysenck (1988) is clear about what he takes the evoked-potential data to imply. Synaptic transmission efficiency, whether it be measured indirectly by RT and IT or more directly by evoked potentials, would be the current contender as the parameter underlying what he has called intelligence A. This is the biological underpinning of intelligence (*g* in Eysenck's terminology), which is totally heritable and uninfluenced by cultural factors. In order to understand intelligence B – that is, the manifestation of intelligence A in cognitive processes, which are to some extent influenced by environmental factors – Eysenck argues that we must first develop comprehensive theories of intelligence A. Therefore, to understand individual differences in intelligence, we must first ask questions such as: What neuro-transmitter gives rise to the variance in synaptic efficiency? or Which type of neurons in the transmission chain contributes most to the variance? and so on. Only when we have explored these kinds of issues sufficiently and have developed a stable biological theory can we begin to investigate the relationship with intelligence B – that is, with cognitive processes.

Eysenck's formulation strikes at the very heart of the question of how intelligence is best understood. This may be a different question to what it is that *causes* individual differences in intelligence. If Eysenck's formulation of the relationship between the AEP and IQ differences is anywhere near the mark, then he may be on the trail of the latter question. However, as far as the agenda set out in chapter 1 is concerned, this knowledge may be of only limited use. We may be interested to discover that variations in neuronal transmission fidelity cause large individual differences in intelligence (although, admittedly, many cognitive psychologists would be horrified at the very idea); but we want to know why this should result in the characteristic cognitive differences found. Obviously a theory based on neural functioning would be very useful for explaining why differences in intelligence are general to many cognitive domains (after all, this is the question that it was designed to answer); but would this help us solve the series of conundrums that we posed as provocative for any theory of intelligence? Why is it that individuals of low intelligence can, nevertheless, outperform average people on some specific cognitive tasks? Why is it that while synaptic efficiency (if it turns out that this is the basis of general intelligence) predicts how well *some* children read, others with high synaptic efficiency have difficulty with reading but not with mathematics? And, crucially, why is it that some very complex computational functions with a substantial cortical basis (such as three-dimensional perception) are apparently unrelated to IQ and yet other apparently simple computational functions, like mental arithmetic, are substantially related to general intelligence? Why is it that the ability to acquire language is within the power of all of us, but beyond the power of the most sophisticated computer, whose silicon speed and fidelity make our neurons look positively slothful in comparison?

Is it the case that before we try to answer such questions we must first sort out the particulars of the physiological basis of general intelligence? Clearly not. To understand these kinds of questions, we need to understand the processing mechanisms that underlie these functions. We may take as our starting-point the hypothesis that there may be differences in the speed, or efficiency, of low-level cognitive processes (which may be biological in origin) which are related to high-level differences in knowledge. The question now is: Why should particular types of knowledge be affected in different ways by differences in general intelligence? The answer to this question requires a theory of the structure of knowledge, a theory of what constitutes general intelligence, *and* (as will become apparent soon) a theory of the *development* of knowledge. These different aspects of a theory of *intelligence* must be able to talk to each other. In order to do this, they must speak the same language. It may be possible to translate what are, at heart, physiological variables into a cognitive language (synaptic efficiency becomes speed of information processing), but as yet no one has found a way of translating higher mental processes into the language of physiology. The crucial distinctions between our concepts of thought simply do not map onto distinctions at a lower level (Fodor 1975, Mehler *et al.* 1984). It is, therefore, a matter of complete indifference for a mechanistic theory of intelligent thought whether intelligence A is based, for example, on glucose uptake or on variations in the sodium pump. Knowing which it is would not help us to answer any of our conundrums.

Some Conclusions from RTs, ITs, and AEPs

We have seen that there has been growing evidence from studies of reaction time, inspection time, and evoked potentials that tasks which appear to be relatively knowledge-free and which tap some low-level aspect of information processing are significantly related to the high-level differences in cognition measured by psychometric tests of intelligence. But it has become clear that these data are by no means as unequivocal as might at first appear. The ingenuity of subjects to discover processing strategies – and indeed, of cognitive psychologists to imagine what these might be – may yet prove that these correlations are based not on the efficiency of knowledge-free mechanisms but on superior processing strategies or on some other, non-cognitive factor. Although this is not impossible, I believe it is unlikely.

It should be clear that the objections to the speed of processing (low-level) hypothesis are unsystematic, differing with each task, and are themselves unlikely to offer alternative explanations of the correlation with the diverse forms of cognitive abilities that these tasks involve. For the strategic hypothesis to be a viable alternative it must be able to explain, *without* appealing to the notion of general intelligence, why, for example, subjects who have higher scores on vocabulary tests are more likely to develop more efficient speed–accuracy trade-off functions in a reaction-time task.

The objections to the speed of processing hypothesis often make appeal to the 'white raven' speculation that philosophers of science are so fond of decrying.[5] That is, they appeal to an as yet unknown factor that may ultimately offer an alternative account of the correlation. So it may be that subject compliance (Mackintosh 1986) or perhaps motivation

---

[5] The statement that there is a white raven (despite all evidence that ravens are black) cannot be disproved, and consequently is scientifically meaningless. The scientific form of the statement is that there are no white ravens.

(Howe 1990) will provide the missing link between performance on these tasks and psychometric intelligence. But as yet, no data are available to support any such speculation. The lack of data will not, I suspect, strain the ingenuity of the critics in thinking up alternative possibilities, as each alternative explanation is discounted in turn.

The most important conclusion is that the most convincing explanation for the relationship will be one embedded in a more general theory of the nature of intelligence. A motivational, strategic, or any factor X explanation will lose out, ultimately, to one which supposes the relationship to be based on a shared reliance on the efficiency of knowledge-free information processing *if* such a relationship can be given increased explanatory value in a wider theoretical structure. In the rest of the book I will try to give the notion of knowledge-free processing this wider theoretical purchase by incorporating it in a more general theory of intelligence.

The First Component of the Model

We came to this chapter suspending our disbelief in general intelligence. We now have some basis, independently of psychometry, for believing that general intelligence is a particular attribute of processing. The data discussed in this chapter converge on the idea that there are low-level cognitive processes that underlie intelligent thinking. Let me gather these processes together and locate them in a particular mechanism: the *basic processing mechanism*. The theory to be developed in this book posits that a major component of intelligence is the basic processing mechanism and that it is responsible for the phenomenon of psychometric *g*, because it varies in its speed among individuals in the population.

The primary importance of this chapter is not the conclusion that there are low-level correlates of intelligence. Rather, the emerging imperative is that our theory be about the relationship between the basic processing mechanism and *knowledge*. To capture the central fact that speed of processing, a knowledge-free processing parameter, is correlated with knowledge-rich performance, I will argue that the basic processing mechanism is responsible for *implementing* thinking. Thinking, in turn, generates knowledge. The task for our theory is to understand the important parameters of thinking that determine why some kinds of thoughts can be implemented, while others cannot. It will turn out that a proper understanding of this relationship is impossible without a developmental dimension to the theory. Before we turn to

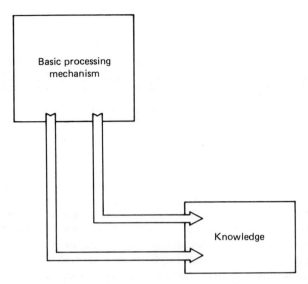

Figure 3.4 *The basic processing mechanism varies in its speed, constrains thought (which generates knowledge), and is the basis of general intelligence.*

this, however, I must redress the balance in my argument. I must justify the proposition that there is, indeed, more to intelligence than the speed, or efficiency, of the basic processing mechanism. In fact, I must now defend the view that the notion of knowledge, or knowledge systems, is of any importance at all for understanding human intelligence.

4

Multiple Minds and Multiple Confusions

The first component of the model is now in place. By postulating a *basic processing mechanism*, which constrains knowledge and thought, we can explain why some knowledge-free information-processing measures correlate with knowledge-rich intelligence tests and why there is a large general component to intelligence test performance. In this chapter I will begin to outline more precisely the relationship between the basic processing mechanism and knowledge. Specifying the relationship between the basic processing mechanism and the mechanisms that generate knowledge is fundamental to a theory of intelligence.

The basic fact discovered by intelligence tests is that individual differences in knowledge are correlated across a wide variety of knowledge domains. As we have seen, this fact is the cornerstone of theories of general intelligence. Yet, if we look more closely at knowledge and its relationship to individual differences in intelligence (and, therefore, *ex hypothesi*, to the speed of the basic processing mechanism), we see that some knowledge is independent of differences in intelligence. It seems, then, that knowledge acquisition mechanisms come in two distinct kinds: those that are constrained by the basic processing mechanism and those that are not. The principal conclusion of the next two chapters is that only knowledge which is acquired through *thought* is related to individual differences in intelligence.

But let me begin by trying to establish that there are, indeed, two kinds of knowledge-acquisition mechanism. This will involve a number of steps:

1 demonstrating that 'intelligence' in humans and computers can refer to quite different kinds of processing mechanisms,
2 demonstrating that the distinction between these kinds of mechanisms is a function of the kinds of information they handle,

3   arguing that these mechanisms map onto a distinction between human knowledge-acquisition mechanisms that is consistent with at least one major theory in cognitive science (Fodor 1983).

Once I have established that it is at least plausible that human knowledge is acquired by means of two different kinds of acquisition mechanism, I then want to show how the distinction increases the explanatory power of the theory by:

1   showing how the distinction between these mechanisms can be used to solve some of the major conundrums of intelligence and development posed in chapter 1,
2   showing that it makes accurate predictions about the nature of specific cognitive functions.

In fact, this distinction will allow a reworking of the notion of specificity that is, paradoxically, consistent with the notion of general intelligence advocated in this book, but inconsistent with a theory of multiple independent intelligences (Gardner 1983).

Intelligence as the Product of a Computational Architecture

The reader may be becoming aware that I have slipped a particular view of psychology in by the back door. I propose that the appropriate level of theorizing for understanding human intelligence is one which views the brain as a collection of information-processing mechanisms designed to furnish us with information about the world. This *computational* perspective offers a new approach to understanding intelligence.[1] If the brain can be regarded as a machine for processing information, then we should be able to make other kinds of information processing machines, like computers, 'intelligent'. The question becomes: What is the *minimal* computational architecture of an information-processing system which could account for the features of intelligence and development that we wish to explain? We have already established one component of the architecture: the basic processing mechanism. What are the others?

---

[1] 'Computational' will be used, in this chapter, in the broadest sense of the term: namely, to refer to processes that could be carried out by a computer. Therefore, a complex computational process is one which it would be difficult to program a computer to do, and a simple computational process is one which it would be easy to program a computer to do.

If we look at the kinds of computational mechanisms usually considered to be central to models of artificial (computer) intelligence, we find a curious paradox: the cognitive functions that seem to embody intelligence for computers and humans are not only different, but mutually *exclusive*. For computers, cognitive problems requiring intelligence for their solution are usually the kind that never feature on intelligence tests. Examples are: extracting three-dimensional representations from two-dimensional stimulus arrays, extracting the linguistic input from a stream of speech, and extracting the syntactic structure from a linguistic input. These are problems of amazing computational complexity; yet these kinds of problems never appear on intelligence tests simply because just about everyone can do them. On the other hand, some tasks that are computationally simple, such as mathematical calculations or remembering lists of symbols, are the very basis of some human intelligence tests. So the paradox is this. Even the mentally retarded, far less someone of just-below-average IQ, have no difficulty in perceiving a three-dimensional world when presented with a two-dimensional retinal image, but no computer yet devised has this kind of capability. However, for some extremely simple computational functions, like those underlying 'mental arithmetic', many of us could not compete with our own digital wrist-watches.

The contrast between what is difficult for a computer and what is difficult for a human being suggests that the computational mechanisms that support human cognitive functions come in two distinct kinds: some that appear computationally simple but show large individual differences and others that are computationally complex but seem to be invariant for members of the species. It is easy to see why this might be so.

Evolution and Information

The kinds of cognitive functions that are computationally complex would also seem to have a plausible evolutionary history. We must be able to differentiate a cliff edge from a lakeside, a predator from our mother, and utterances such as 'Lions are eager to kill' from 'Lions are easy to kill'. It is worth (in evolutionary cash value) building complex processing mechanisms, dedicated to particular functions, if at least three conditions hold: (1) the information they provide increases the organism's chances of survival; (2) the information base for those mechanisms is likely to be invariant for members of the species; (3) the information cannot be provided by more general-purpose knowledge-acquisition mechanisms in an ecologically useful time frame. A cognitive function, such as ecological visual perception (Gibson 1979), easily meets those requirements.

We can presume that the proximal stimulus for visual information-processing is, and has always been, the same for all human beings. It seems reasonable, then, to suppose that nature has endowed all of us with a processing mechanism that gives us the *capability* of extracting the three-dimensional structure from the two-dimensional stimulus array. Certainly processing mechanisms that could deliver this information would be at an evolutionary premium, and, as the attempts to develop computerized visual systems make clear, it is hard to imagine anything other than a dedicated visual processing system that could extract this information fast enough for it to be useful.

Contrast this with the situation in which we need to acquire knowledge that may be idiosyncratic. This kind of knowledge has to be extracted from a highly variable information base. For example, whether it will be important to know about the edibility of different kinds of tropical fruits, the construction qualities of different kinds of snow, or the motivations of middle management will vary with socio-cultural circumstances, and no two humans beings will have identical opportunities for acquiring this kind of information.

Given this difference in the kind of knowledge to be acquired, it would seem reasonable to suppose that these knowledge systems are subserved by different kinds of mechanism. It is unlikely that a *general-purpose* knowledge-acquisition device would have the capability of learning the syntax of a language (Chomsky 1986). Equally, it is hard to see how the processing mechanisms postulated by Marr (1982) could be used for anything other than constructing a $2\frac{1}{2}$-dimensional sketch from retinal information. So, on the one hand, there are those mechanisms that subserve idiosyncratic knowledge acquisition, which will be general-purpose computational mechanisms, whose operations will be constrained by the speed of the basic processing mechanism.[2] By contrast, there are those knowledge-acquisition mechanisms designed to provide us with particular kinds of information of evolutionary importance. I will call the latter 'modules', after the notion developed by Fodor (1983). Their defining characteristics in my theory are that they are functionally independent, complex, computational processes of significant evolutionary importance, which show no relationship to individual differences in intelligence. Since we have no reason to believe that modules show variance associated with individual differences in

---

[2] These knowledge-acquisition mechanisms will themselves show individual differences, as we shall see in the next chapter.

measured intelligence, this must mean that their computations are unconstrained by the basic processing mechanism. Evolution will have ensured that such important information is provided to all undamaged members of the species.

The proposition that modules are unconstrained by the speed of the basic processing mechanism helps to resolve some of the conundrums posed in chapter 1. For example, if ecological visual perception is afforded by a dedicated processing module, then this would explain why the mentally retarded seem perfectly capable of perceiving the visual world. Simply put, this process is not implemented via the basic processing mechanism, and is therefore unrelated to individual differences in intelligence. There again, if certain cognitive processes involve the operation of processing modules, then individuals who have a damaged module may perform much worse than would be expected from their level of general intelligence (which is more an indication of the speed of the basic processing mechanism). For example, autistic children of equal mental ages to Down's syndrome children cannot seem to solve problems which involve attributing mental states to others, whereas Down's syndrome children have no such difficulty (Baron-Cohen *et al.* 1985; Frith 1989). Perhaps, then, core computational capacities involved in a 'theory of mind' (Leslie 1987) emerge independently of level of intelligence (otherwise it is hard to understand how low-IQ Down's syndrome children can solve problems involving such complex concepts as 'false belief') but may be selectively impaired by brain damage. We shall return to the case of autism again in chapter 8.

The key feature of modules is that their computations are unconstrained by the speed of the basic processing mechanism. This feature can also help to clear up the popular misconception that theories of modularity are incompatible with the notion of general intelligence and favour instead the view that intelligence consists of a collection of independent abilities (Gardner 1983). General intelligence derives from the constraint that speed of processing imposes on thinking, and, for at least one influential theory of modularity, the processes underlying thinking and the processes underlying modular functions are quite different (Fodor 1983).

Modularity of Mind

Fodor's monograph entitled *The Modularity of Mind* provides a detailed argument for supposing that there are two different kinds of cognitive systems: *input systems*, which are modular and independent of one another, and *central processes*, which are concerned with the fixation of belief.

Fodor argues that the function of input systems is to provide central systems of thought with representations, based on the processing of stimulus information, of the current state of the world; for example, that there is a rather hungry-looking lion well within leaping distance. The major point is that if there is a predator in the locale, then it is useful to know about this *fast*. The degree of complexity of the computational process involved in extracting such information from the retinal array is such that if we had to rely on acquiring the information using the problem-solving techniques common to central mechanisms of thought ('Hey, there is a large object in my visual field. Is it a bus or a lion or a smear on my sunglasses?'), we would probably be eaten before we had tested our first perceptual hypothesis. It is in central processes that the information afforded by the module is 'evaluated'; and this evaluation will be sensitive to the state of the entire cognitive system. That is, what we believe or think at any one moment can be influenced by a potentially infinite number of cognitive states. Thus, although in the particular example cited there will probably be little pause for cogitation, my belief about the world, on being presented with this information, will ultimately depend on whether I believe that there are lions roaming the streets of the city, a belief which, in turn, may depend on whether I think the circus is in town or whether I know it is student rag week or whether I believe I have been overworking, and so on, *ad infinitum*. This evaluative function of central processes (thought) makes them slow, context-sensitive, highly malleable, and, ultimately, indeterminate.

On the other hand, the *input systems* (modules), which present central processes with representations of the current state of the world, have exactly the opposite characteristics. They are domain-specific, fast, hard-wired, mandatory (when presented with the appropriate stimulus, they will respond by outputting an appropriate representation to central systems), and informationally encapsulated. They are uninfluenced by whether or not it is student rag week; they 'know' a lion when they see one (and a good job too). So, given an appropriate stimulus in their specialized domains, they will respond speedily and reflexively, by supplying central processes with a representation of the stimulus event. These key features of input systems mean that the internal operation of each Fodorian module is opaque to the others. Given a high-fidelity processing mechanism that rarely makes mistakes, an evolutionary strategy of 'run now and think later' makes sense.

On Fodor's analysis, many aspects of cognitive functioning – for example, perception (both auditory and visual) and language (including syntactic parsing, encoding of speech, and so forth – are underlain by quite different kinds of processes from those underlying thinking. Further, Fodor explicitly recognizes that 'intelligence' is a property of thought, since he regards modules as essentially complex, but 'dumb', mechanisms. The distinction between Fodor's modules and his central processes of thought parallels the distinction, argued for earlier in the chapter, between knowledge-acquisition mechanisms that are computationally complex but show no individual differences and simple computational processes involved in thinking and reasoning in which there are large individual differences. Where I differ from Fodor, however, is in positing that processes of thought are also constrained, principally by the basic processing mechanism.

Since it is hypothesized that general intelligence depends, primarily, on the speed of the basic processing mechanism, and this, in turn, is a mechanism that implements thinking (a Fodorian central process), then there is no conflict between this version of general intelligence and Fodor's theory of modularity. Simply put, modules are not examples of what is usually meant by 'multiple intelligences'. To see why modules are not multiple intelligences and why the notion of multiple intelligences is untenable, let us take a closer look at one particular theory.

Multiple Intelligences

The best-known of current theories of multiple intelligences is that of Gardner (1983). His central proposition is that there are a number of clearly separate intelligences, each specialized for a different cognitive domain. This proposition is at odds with all the psychometric evidence for general intelligence amassed in chapter 2 (and with the cognitive evidence presented in chapter 3). This does not cause any particular problem for Gardner, however, since he takes the view that intelligence tests, and particularly the notion of general intelligence which they have spawned, have been harmful for the scientific study of individual differences in intelligence. Psychometric tests, he argues, have focused on only a limited aspect of intelligence – namely, the logical-mathematical and verbal skills so valued by Western culture. Therefore we should not be surprised to find that these abilities 'go together'. If Gardner dismisses the evidence from intelligence test data, which, as we have seen, runs strongly against the notion of multiple intelligences, what evidence does he find to support it?

Gardner's strategy is to look for 'signs', or 'markers', that might indicate whether a cognitive ability constitutes an 'intelligence'. The most prominent of these signs are examples of extreme ability, core cognitive (symbolic) operations, developmental profile, brain localization, evolutionary history, and, finally, cultural significance (they should not merely reflect performance on 'meaningless' psychometric or laboratory tasks). So, for example, Gardner argues that there is a linguistic intelligence, since we can find cases of extreme ability (poets and writers); there are core cognitive operations (syntax, semantics, pragmatics); there is a developmental story (babies cannot talk) and some evidence of brain localization (Broca's and Wernicke's areas); there is a plausible evolutionary story (we are the only animals that display linguistic intelligence, and yet, he argues, components of linguistic intelligence are to be found in bird-song and primate communication); and, finally, linguistic intelligence has important cultural impact. Gardner claims that the list of signs is incomplete and that, further, a candidate intelligence need not meet all the criteria. After looking for these kinds of signs in an impressive search of biology, cognitive psychology, and social anthropology, Gardner settles on six candidate intelligences: linguistic, musical, logical-mathematical, spatial, personal, and bodily-kinaesthetic.

A closer look makes clear why this catholicity of approach is a hindrance, rather than an aid, to a coherent theory of intelligence. As used by Gardner (1983), an intelligence has no theoretical status. It is sometimes a behaviour, sometimes a cognitive process, and sometimes a structure in the brain. This enables evidence from brains, cognition, and culture to be used as supportive evidence for autonomy, with the only constraint being that the brain structure or the cognitive process or the culturally valued behaviour *looks like* it has something to do with 'language' or 'music' or whatever the intelligence under investigation is. This is despite the absence of any theoretical connection between the brain structure, the cognitive process and the culturally valued behaviour. For example, if we look at what Gardner calls *linguistic intelligence*, we see that the *biological* support for its autonomy stems largely from studies of the consequence for language use of damage to certain areas of the brain. Shift a level, and we see that autonomy is claimed for the core *cognitive* operations of syntax, semantics, and pragmatics. Shift a further level, and we find support for autonomy in the fact that we find individuals considered to be great poets or great musicians but rarely both. Nowhere is it stated whether linguistic intelligence is considered to be a cultural phenomenon with biological and cognitive underpinnings or a biological structure with cognitive and cultural manifestations or a cognitive system with a biological underpinning and a cultural manifestation. This allows the necessary condition, of demonstrating that these independent strands of evidence map onto the same 'thing', to be avoided. Leaving aside the quality of the evidence for autonomy within each level of analysis, we simply have no way of telling whether 'linguistic intelligence' refers to the same construct at each level. Do the autonomous brain structures map onto the autonomous psychological processes of syntax, semantics, and pragmatics, and do these functions, in turn, map on to the culturally valued systems of thought by which we can distinguish Shakespeare from less gifted scribes? Evidence for autonomy is acceptable only where either the evidence is at the same level of description (biological, cognitive, cultural) as the 'intelligence' or where there is a separate theory which states the relationship between the different levels of description and what that implies for autonomy. Since these conditions do not hold, we can conclude that Gardner's evidence for multiple intelligences is illusory. Rather than the evidence from a diversity of sources being used as the scaffolding within which a theory is constructed, it turns out that the scaffolding *is* the theory.

But perhaps we should try a different tack to avoid these shifts in level. Could we rescue the notion of multiple intelligences by choosing one level of description for an intelligence? If Gardner were willing to equate an intelligence with a computational mechanism, could he not argue that the evidence for the independence of processing modules presented earlier in this chapter has vindicated his theory? Herein lies the source of the confusion.

As we have seen, there are good reasons for supposing that there are autonomous cognitive processes. However, it should be clear that the kind of multiple minds implicit in, say, a theory of modularity is not coextensive with multiple intelligences. A look at the properties of Gardner's intelligences and Fodor's modules confirms that they are not so much inconsistent as mutually exclusive. Modules are fast; intelligences are slow. Modules are mandatory; intelligences are optional. Modules are hard-wired; intelligences are malleable. Modules are dumb (reflexive); intelligences are bright. Modules are input systems; intelligences are remote from peripheral input devices. The operating characteristics of Gardner's intelligences and the domain of knowledge to which they are relevant make multiple intelligences a theory of central processes. Fodor's modularity thesis, by contrast, applies to input systems, *not* central processes. The debate about whether intelligence is general or a collection of independent abilities is, then, a debate about whether central processes, or thought, have general or multiple properties. Autonomy of processes *per se* does not necessarily bear on the issue.

In this chapter it has become clear that the appropriate level of description for the theoretical construct 'intelligence' is a cognitive or computational one (for this seems to map onto the phenomena of interest most readily). We can now use this more specific definition of the term 'intelligence' to evaluate whether 'intelligence' is multiple or general. If we count those cognitive functions served by complex processing modules as aspects of intelligence, then we would undoubtedly have multiple intelligences. But when we consider central processes of thought, the notion of multiple, independent thinking abilities (intelligences) is clearly untenable. These processes have much in common. It should now be obvious that the theory of multiple intelligences has managed to survive obvious empirical refutation only by appealing to argument and evidence pertinent to quite different kinds of cognitive processes from those involved in *thinking*.

Specificity Revisited

In chapter 3 I argued that knowledge will be highly constrained by the speed of the basic processing mechanism, which means that individual differences in knowledge will reflect the property known as general intelligence. In this chapter I have developed the argument by proposing that there are some knowledge mechanisms that are free from the constraint of the basic processing mechanism. This allows scope for specificity of function within the theory; but the specificity does not apply to the domain of intelligent thought, and so the theory cannot be regarded as a variant of existing theories of multiple intelligences. This new perspective allows us to reexamine other evidence for specificity, which may have been mistakenly interpreted as evidence against the notion of general intelligence. Thus, the theory predicts that evidence for specificity should be strongest for modules – that is, for complex computational processes of significant evolutionary importance which show no relationship to individual differences in intelligence. We can now turn to neuropsychological data to provide some test of this hypothesis.

Neuropsychology and Specific Abilities

Specific deficits

The most compelling data on the relationship between the brain and intellectual functioning come from the study of brain-injured patients.

The first thing to note is that there is not a great deal of evidence, in any case, in favour of the notion of specificity of function. It is rarely the case that brain damage results in specific, rather than general, cognitive deficits. Further, the evidence for a one-to-one relationship between particular kinds of lesions (their location, extent, etiology) and particular psychological deficits is also weak. This is only to be expected, of course. It would be unreasonable to suppose that a blow from a hammer or the explosion of a cerebral artery would respect functional boundaries. In addition, it is usually difficult to be sure about the extent and severity of a lesion, since effects may vary with many idiosyncratic characteristics of patients.

Despite the difficulty in obtaining unequivocal evidence, the prevailing *Zeitgeist* is in favour of specificity.[3] There are two principal reasons for this: first, neuropsychological models are increasingly being driven by cognitive models (Ellis and Young 1988) which, as we have seen already, have an inherent bias towards specificity of function; second, there are a few examples of extreme specificity, which, when they occur, are spectacular. It is in these examples that we might expect to find evidence for specificity based on dysfunction in modules.

The first indication that specificity might indicate dysfunction in modules comes from the nature of the instruments used for assessment in neuropsychology. Standard intelligence tests are usually regarded as too blunt to be useful; that is, they do not provide measures of the functions that clinical practice indicates are implicated in specific deficits.[4] As a result, neuropsychologists have tended to develop their own set of diagnostic tools, using many tests that do not appear in a standard intelligence test battery. For example, the Luria–Nebraska test battery contains tests of motor functions, pitch perception, and tactile perception; while the Halstead–Reitan battery also tests such functions as grip-strength, speech-sound perception, and, again, tactile perception. This suggests that *specific* deficits may be the province of, classically, non-intellectual domains.

This line of argument is consistent with the often spectacular data from clinical case studies. Sacks (1985) emphasizes in his collection of mysterious cases that specific deficits seem bizarre only because they

---

[3] It will be interesting to see if Lashley's notions of mass action make a comeback, given the growing interest in cognitive science in parallel distributed processing.

[4] The only psychometric distinction which seems to have any influence on neuropsychological models is that between the broad categories of verbal and spatial intelligence. For example, differences between Verbal and Performance scores on the WAIS are frequently quoted when dysfunction of one of the cerebral hemispheres is suspected. This exception is important, as we shall see in the next chapter.

occur in systems that normally work so well that their functioning passes unnoticed. That is, specific deficits occur in psychological systems that normally show no variance. Thus, the most compelling evidence for specificity is seen in bizarre patients such as prosopagnosics, who have a deficit in the ability to recognize faces (De Renzi 1986); deep dysphasics, who mistake the 'meaning' of words in a bizarre fashion (Marshall and Newcombe 1966, Morton and Patterson 1980); and visual agnosics, for whom the recognition of everyday objects is extremely difficult (Humphreys and Riddoch 1984). However, even in these cases the degree or domain of specificity is not always clear. For example, in cases of prosopagnosia the recognition deficit for faces is often accompanied by failure to recognize other familiar objects (Newcombe and Young 1989). Furthermore, there are disputes as to what stages of processing are implicated in the primary deficit in brain-damaged patients – for example, whether the deficit lies in perceptual or in retrieval mechanisms.[5] Nevertheless, what is clear is that the nature of the specificity shown in these clinical cases is not that normally captured by tests of specific abilities on intelligence tests.

The fact that specific deficits occur for functions that are not tested by traditional intelligence tests and that these functions also appear to be computationally complex suggests that specific deficits may be attributable to damage in the mechanisms I have called 'modules'. Conversely, thought processes are not subject to specific deficits, even though, on occasion, the products of *previous* thoughts may be subject to selective damage.

It is obvious that much *encyclopaedic* knowledge (for example, knowing that a strawberry is a fruit) is acquired through thought and, further, that such knowledge is the very stuff of the information-based subtests of many intelligence tests. There are cases of brain damage which results in a specific deficit in the recognition and naming of semantic categories, such as *living things* or *foods* (Warrington and Shallice 1984; Warrington and McCarthy 1987). Do such cases violate the claim that when brain damage influences thought, the effect is general rather than specific? I believe not. It seems likely that these cases represent instances in which the *products* of previous thoughts (or crystallized intelligence in the framework of Horn and Cattell (1966)) are selectively impaired by brain damage. Thus, it is telling that the semantic categories demarcated by brain damage do not correspond to any distinctions between psychometric abilities. For example, no test makes

---

[5] See Ellis *et al.* (eds) 1986 for discussions of prosopagnosia and Shallice 1988 for a discussion of dyslexia and object recognition.

a principled distinction between the knowledge that a strawberry is a fruit and the knowledge that a chair is a piece of furniture (except, of course, in terms of their relative *difficulties*, which are, by and large, the same for everyone). In conclusion, while there may be selective damage to the *modular organization* of knowledge, rarely, if at all, do we find case studies of brain-damaged patients who have lost a single Primary Mental Ability, as might be measured by Thurstone's intelligence test battery.[6] When 'thinking' abilities are disturbed by brain damage, the effect is general, rather than specific.

Isolated abilities

What of the opposite case where, rather than finding a specific deficit, we find the sparing of an isolated function? This is seen when an individual of very low measured intelligence displays a remarkable isolated ability. Such cases have become known as *idiots savants* (O'Connor and Hermelin 1984). For example, Cromer (in press) reports the case of D.H., who has spina bifida and displays features of what has been described as 'cocktail party syndrome' (Hadenius *et al.* 1962). D.H. has very low measured intelligence (she cannot count, read simple words, or write her own name, and does not know the order of the seasons and so on), yet has extremely good conversational skills. She can engage in an extremely verbose conversation, obeying all the rules of syntax, generating utterances which are semantically coherent, and with conversational pragmatics of a highly developed nature; yet, when tested, she doesn't seem to understand the meaning of the majority of the words or utterances she uses. This bizarre isolation of a linguistic function is, generally, in line with the modular hypothesis, in which the language 'faculty' (Chomsky 1986) is often regarded as a module *par excellence*.[7] In the case of other examples of *idiots savants* (calendrical calculators,

---

[6] One possible exception to this may be rare cases of dyscalculia, which is a special deficit in numerical ability (Warrington 1982). But again, it is not clear whether the damage is to a *store* of acquired knowledge or the ability to implement calculating procedures (although see Temple 1991).

[7] In this book I will often talk of language as if it were a single module. This, of course, is simply a convenient shorthand, in that 'language' is a *system* that is composed of many modules. Indeed, the theory of the minimal cognitive architecture acknowledges this by using both phonological and syntactic processing as examples of modules. Apart from being a convenient shorthand, envisaging language as modular captures the general proposition that many language processes are independent of central cognitive processes. It also avoids the necessity of taking a stance on the many disputes about which processes contained in the language module are innately specified and which result from a developmental process of modularization.

artists and musicians) the evidence for specificity being confined to modules is not so clear. The phenomenon of *idiots savants* will be discussed again in chapter 8 but, for the moment, suffice it to say that no *idiot savant* displays an isolated ability that would be recognizable as a subscale from the WAIS.

Evidence from the localization (as distinct from the independence) of function is also broadly consistent with the module hypothesis; that is, if any function can be said to show any degree of localization, that function will look like a module in my theory (to repeat: a complex computational process of significant evolutionary importance which shows no relationship to individual differences in intelligence). There is very little unambiguous evidence for the cortical localization of higher cognitive functions, whether from brain-damaged patients, experimental observations of intact humans, or animal studies; but the most robust evidence seems to be for cortical localization of some visual and language processing. Thus, it is clear that the striate cortex is specialized for visual processing and that Broca's and Wernicke's areas in the left hemisphere are implicated in linguistic comprehension and production. Both vision and language have already been singled out as likely exemplars of cognitive functions supported by modules. Attempts to associate other cortical areas with other higher cognitive functions that may be associated more with thinking than with the functioning of modules have, in general, failed. These attempts have usually resorted to ill-defined constructs, such as 'planning' ability (frontal lobes), which themselves have a dubious status as psychological processes.

The point about the neuropsychological evidence on specificity of function is that, rather than being incompatible with the theory advanced in this book, it is remarkably consistent with it. If thinking abilities are affected by brain damage, it seems that the effect is general. Where specificity is found, it seems to be in those very functions that would be primary candidates for modules.

Modules as the Second Kind of Processing Mechanism

We have now established a second type of processing mechanism in the minimal cognitive architecture that can account for the data on intelligence and development (see figure 4.1). Modules are complex computational functions that have been shaped by evolutionary pressure to provide us with information that would be impossible to obtain using general-purpose problem- solvers. The information provided by modules is so important for our survival that all undamaged human

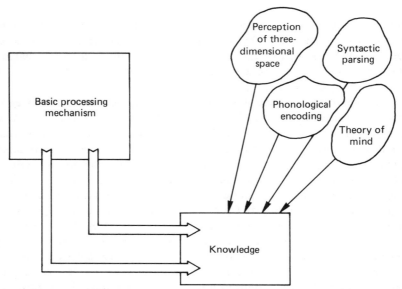

Figure 4.1 Modules provide knowledge unconstrained by the speed of the basic processing mechanism.

beings, irrespective of levels of measured intelligence, have access to it.

Modules can accommodate much of the evidence on specificity of cognitive processes that previously may have been taken as evidence against any theory which takes general intelligence seriously. In addition, they can explain some of our important conundrums: that some complex cognitive functions are within the capability of even the mentally retarded, that some complex abilities appear in isolation, and that specific cognitive deficits are found in individuals of otherwise normal intelligence.

There is, of course, much current debate in cognitive science about the computational characteristics and the domains of operations of processing modules (Karmiloff-Smith 1986; Leslie 1986, 1987; Marshall 1984; Spelke 1987). We will return to these issues time and again throughout the rest of the book. Because specifying which psychological processes should be considered as modules in my theory is less important than establishing the validity of the notion, I am loath to speculate on what they may be, for fear that an argument over the merits of any particular exemplar may obscure the major thrust of the construct. Nevertheless, it should be clear that the major likely examples are modules which support ecological visual perception (Marr 1982), language acquisition (Chomsky 1986), and many aspects of speech perception, all of which have been argued for by Fodor (1983). Such modules are, by and large, input systems. In addition, it seems likely that many of the phenomena of psychological functioning seen by developmental psychologists as central to the development of intelligence, such as causality (Leslie 1986, Das Gupta and Anderson, in prep.), the object concept (Baillargeon 1986), perceptual categories (Mandler 1988), and 'theory of mind' (Leslie 1987), may be supported by processing modules. Such modules may have important differences from the ones that could be characterized as input systems. For example, in Leslie's theory the modular core of 'theory of mind' is contained in a single function – a *decoupling mechanism* – which allows children to represent representations (metarepresentation). Although such a function may be at the opposite pole on a computational complexity dimension from, for example, a module underlying ecological perception, it still has modular status in the theory of the minimal cognitive architecture, because its operation is unconstrained by the speed of the basic processing mechanism. But perhaps it is time to formally recognize that modules in the minimal cognitive architecture may look like a heterogeneous collection of mechanisms when viewed through computational lenses.

Modules Mark II?

The discussion in this chapter has centred on modules as computationally *complex* processing devices that are probably innately specified. I will now refer to these modules as 'modules Mark I'. I have concentrated on Mark I modules to highlight the distinction between complexity of thought, which is constrained by the speed of the basic processing mechanism, and complexity of processing, which need not be so constrained if the complex process is modular. However, the defining feature of a module in the minimal cognitive architecture is that its computations are not implemented via the basic processing mechanism. This opens up the possibility of another class of processing mechanisms whose operation is unconstrained by the basic processing mechanism. I am thinking of two kinds of related mechanisms which, unlike Mark I modules, are not, in the first case, computationally complex or are not, in the second case, innately specified. I will call these mechanisms or processes 'modules, Mark II'.

The first kind of mechanisms that may be Mark II modules are the fetch-and-carry operations that make information-processing possible. Such mechanisms may constitute part of what Pylyshyn has called the 'functional architecture' of cognition (Pylyshyn 1980). For example, information-processing systems must have mechanisms that allow information to be moved to and from current processing and memory. These mechanisms are not computationally complex, but elemental, or primitive, for any information-processing system; for without them the system could not operate normally. It is therefore significant – and yet, perhaps, surprising – that many of the mechanisms that underlie long-term memory storage and retrieval are uninfluenced by individual differences in intelligence (N. R. Ellis 1963) and, therefore, would constitute processing modules.[8]

The second kind of processes that may have modular status within the minimal cognitive architecture are those that have become known as automatic in the information-processing framework of Shiffrin and Schneider (1977). Here, due to extensive practice, information-processing routines function independently of thought. There are significant similarities between the notion of module and the idea of automatic processes in addition to their shared feature of automaticity. Modules and automatic processes are fast; they are mandatory; and they are cognitively impenetrable. However, from the perspective of this book, their most important shared characteristic is their immunity from the influence of individual differences. Indeed, Hasher and Zacks (1979, 1984) have as one of their main criteria for automatic processes that they should show little or no effect of age or individual differences. Experiments comparing automatic processing in mentally retarded and non-retarded subjects and in subjects of varying ages have substantiated their claim that these processes meet these criteria (Ellis *et al.* 1988; Nigro and Roak 1987). It seems, then, that automatic processes share the defining feature of Mark I modules. The major difference may be that whereas modules are hard-wired and operate on specified stimulus types, automaticity can be a feature of any arbitrary sequence of information-processing operations given enough practice. It is for this reason that automatic processes should be thought of as modules Mark II. Making this distinction allows for the possibility that acquired knowledge can itself serve the function of a complex processing module which can control information processing in the absence of thought.

Modules and knowledge

Inevitably the interactions between thought (constrained by the basic processing mechanism) and knowledge will be an iterative procedure involving the knowledge-acquisition mechanisms of the minimal cogni-

---

[8]   Surprising, because memory tests feature prominently in intelligence test batteries. Yet these are tests more of the quantity and quality of material that has been stored in long-term memory than of storage and retrieval mechanisms.

tive architecture and knowledge itself. In this chapter I have emphasized the kinds of knowledge it would not be possible to acquire
through thinking – that is, the complex representations afforded by
modules (of the Mark I variety) underlying visual perception, syntactic
and phonological processing, and, perhaps, 'theory of mind'. Equally,
knowledge that is acquired through thought may subsequently assume
modular status provided that its subsequent processing does not, itself,
require thought when used. This could happen, for example, because
the computational process itself might become automatic through
repeated practice. Reading may turn out to be a good example of such
an iterative process. Reading can be shown to involve many modular
processes (Shankweiler and Crain 1986), most of which are unlikely to
be modules in the Mark I sense. However, the process of reading may
involve fetch-and-carry mechanisms which are Mark II modules and
may also depend on phonological capabilities buried deep within a
Mark I module (see the discussion in chapter 8). If any of these processes are damaged – say, the ability to phonologically segment speech
– this may result in a specific developmental reading deficit (at least for
alphabetic scripts (Frith 1986)) that is independent of a child's level of
general intelligence.

Levels of Explanation Revisited

The notion that modules are unrelated to individual differences in
intelligence (that is, are unconstrained by the speed of the basic processing mechanism) is also important *vis-à-vis* a major goal of this
book, which is to establish that cognitive psychology provides the
appropriate theoretical constructs for understanding the relationship
between intelligence and development. Thus, by the end of the previous chapter it was clear that variations in *low-level* processes could
account for many of the differences found in acquired knowledge.
However, the *module/thought* distinction is a computational distinction
(rather than a neurophysiological or a cultural one), and, since we
would fail to understand much of human intelligence without making
this distinction, it follows that a theory of intelligence must be a cognitive theory. Thus, the distinction between processing modules and central processes of thought cleaves the notion of intelligence into two
'natural kinds' and indicates that, contrary to the reductionist agenda
(Eysenck 1986), cognitive processes are not epiphenomena of neurophysiology; nor will a theory of intelligent thought have to wait for an
understanding of the neurophysiology of general intelligence before it
can begin its work. Indeed, the importance of the notion of modules

will become even more obvious when we look at development, particularly when we consider why individual differences in intelligence remain so stable during a period of massive changes in cognitive competence.

But first, there is still one component missing from our minimal computational architecture: the knowledge-acquisition mechanisms that are subject to the constraint of the basic processing mechanism. These are the subject of the next chapter.

5

Knowledge, Specific Abilities, and General Intelligence

The first two components of the theory are now in place. In chapter 3 we saw the need to explain (a) why low-level knowledge-free measures of information processing are good predictors of high-level knowledge-rich cognitive performance, and (b) why intelligence is general, but knowledge specific. The solution was to distinguish between a mechanism underlying knowledge *acquisition* and knowledge itself. The hypothesis is that there is a *basic processing mechanism* that varies in its speed, or efficiency, between individuals in the population. The basic processing mechanism is responsible for implementing *thinking* – that is, the process by which knowledge is acquired. Individuals with faster basic processing mechanisms will be able to acquire more voluminous and complex knowledge than individuals with slower basic processing mechanisms; yet the knowledge that is acquired is, itself, domain-specific. Since general intelligence derives from a property of a mechanism involved in knowledge acquisition (the speed of the basic processing mechanism), rather than a property of knowledge itself, there is no contradiction between general intelligence and domain specificity in knowledge. The implications of this mechanism for a theory of intelligence are clear:

1 individual differences in general intelligence have their basis in a knowledge-free parameter of processing;
2 the focus of the theory now becomes the specification of the nature of the relationship between the basic processing mechanism and knowledge.

Chapter 4 emphasized the importance of the distinction between the basic processing mechanism, as the basis of individual differences in intelligence, and knowledge. It turns out that when we look closely at the kinds of knowledge available to human beings, one kind seems to be immune from the effects of individual differences in intelligence. This must mean, *ex hypothesi*, that the computational procedures used by the mechanisms responsible for the acquisition of this kind of knowledge are not implemented by the basic processing mechanism. I called such mechanisms 'modules'. They are hypothesized to be dedicated, computationally complex mechanisms whose function is to provide us with evolutionarily prescribed information which could not be furnished by general-purpose problem-solvers.

The purpose of the current chapter is to specify the nature of the knowledge-acquisition mechanisms that are, indeed, constrained by the speed of the basic processing mechanism.

Acquiring Idiosyncratic Knowledge

A good place to start the characterization of the mechanisms that are constrained by the basic processing mechanism is to contrast them with the mechanisms that are not so constrained: that is, the modules. There are two main dimensions on which they can be contrasted.

The first concerns *the nature of the knowledge they acquire*. Modules could not have evolved were it not for the fact that the information base for some extremely important knowledge (for example, how surfaces in the environment are arranged, the syntactic structure of language, that other people think/believe/desire) is an evolutionary invariant and is species-universal. However, many have conjectured that what makes human beings truly intelligent is their *adaptability* – that is, the ability to acquire knowledge that is *locally* useful. By this I mean knowledge that is useful in terms of the needs and the environmental circumstances of particular individuals. There must be knowledge-acquisition mechanisms that allow human beings to acquire what are, in evolutionary terms, arbitrary cognitive skills. For example, these mechanisms might need to furnish us with the skill of navigation, perhaps through the observation of stars or through using a compass or by reading a map. We might have to program a computer or distinguish poisonous from non-poisonous fruit or plant crops or write books. Much of what we need to know depends on local environmental and cultural circumstances, and it is this kind of knowledge that is most related to individual differences in intelligence.[1] Thus, one characterization of the

---

[1] Of course, factors other than intelligence are important for the acquisition of this kind of knowledge – for example, practice, opportunity, and motivation.

mechanisms that are constrained by the speed of the basic processing mechanism is that they are *general-purpose* problem-solvers. The history of general-purpose problem-solvers in cognitive science is a chequered one, but for the moment I simply want to contrast the nature of mechanisms whose computational functions are likely to be dedicated to one important task with those mechanisms which must subserve *idiosyncratic* knowledge-acquisition.

The second major contrast between modules and the mechanisms that are constrained by the speed of the basic processing mechanism concerns variability in operation. Modules are all or nothing mechanisms; they work either well or not at all, whereas the constrained mechanisms exhibit individual differences – that is, the quality of their operation is normally distributed in the population. How do we know this? If the mechanisms constrained by the speed of the basic processing mechanism were like modules – that is, if they showed no normally distributed variability in their operation – then they would not contribute to variability in intelligence, and all such variability would be attributable to variation in the speed of the basic processing mechanism. In psychometric terms, this would mean that all variation in abilities would be attributable to differences in general intelligence. However, there is a great deal of evidence to suggest that there are individual differences in more specific abilities. With these abilities in mind, I have called the mechanisms that are constrained by the speed of the basic processing mechanism the 'specific processors'. The questions now are: How many specific processors are there (the most theoretically trivial question)? What is the nature of their computations? And how are they constrained by the speed of processing of the basic processing mechanism?

To answer these questions, we must look to the major areas in psychology concerned with specific cognitive abilities. In particular, psychometrics, neuropsychology, and behaviour genetics provide a great deal of data that can help answer the first question (how many?) and perhaps provide a preliminary answer to the second (what is the nature of their computations?). Indeed, there is no shortage of candidate specific abilities. On the contrary, the problem is to narrow down the many available options. For example, some psychometricians think that there are at least 120 potential specific abilities (Guilford 1966). I hope to narrow the options by focusing on major *commonalities* among these different disciplines. In fact, in the next sections we need only make a superficial examination of these research areas, because the conclusions drawn are, in the main, uncontentious. There is an obvious consensus that there are *at least* two specific abilities, and that these abilities map onto a distinction between verbal and spatial ability. Finally, the notions of verbal and spatial ability will be subject to a more detailed cognitive analysis (and justification of their status as specific abilities), since, as we shall see later, only a cognitive exposition can answer the questions concerning the computational nature of the specific processors and their relationship to the basic processing mechanism.

Specific Processors and Specific Abilities

The Psychometry of Specific Abilities

As we have already seen in chapter 2, psychometric theorists have long supposed that there is more to intelligence than g. Their views have ranged from Spearman's minimal concession that each test also measures a test-specific component to Guilford's contention that there are at least 120 abilities and no g. In between we had Thurstone, who proposed that intelligence was composed of eight primary mental abilities, and Vernon, who acknowledged the primary position of g but also recognized that other group factors, identified as verbal educational (v:ed) and spatial/mechanical (k:m), accounted for significant proportions of the variation in intelligence test scores. Clearly, the psychometric tradition offers a wide range of potential specific abilities, with no unequivocal method of selecting among them. This ambiguity is particularly serious if we are interested in the relationship between general intelligence and specific abilities, which is, *ex hypothesi*, the psychometric equivalent of the basic processing mechanism/specific processor relationship, which is the subject of this chapter.

While we may build in assumptions about what should be considered a reasonable set of cognitive tests (for example, heterogeneity) on which an analysis should be based (Jensen 1987a), if the nature of general intelligence and specific abilities depends on the particular constituents of test batteries, psychometrics can offer us only clues, not answers, to the question of how the intellect is structured.

Nevertheless, no psychometrician would doubt that there are a *minimum* of two distinct abilities, which are equivalent to the major group factors and most obviously labelled 'verbal' and 'spatial' ability. Therefore, psychometric evidence suggests that there are *at least* two *specific processors*. Even though this may be considered as too few by most psychometricians, it is probably the case that there is less unanimity over the validity of the other abilities than there is for the verbal/spatial distinction.

The Neuropsychology of Specific Abilities

In chapter 4 I argued that much of the evidence for *specificity* in psychological functions derived from neuropsychological studies is actually relevant to specific *competencies*, rather than specific *abilities*. That is, the very functions that seem to be prone to specific damage are not to be found in standard measures of individual differences in intelligence, simply because they do not appear to constitute dimensions of individual variation. However, there is one caveat: there seems to be converging evidence from neuropsychological studies that there are at least two specific abilities with normal distribution in the general population. This evidence comes from three main research areas.

(1)   Das *et al.* (1979) identify two distinct processes which can be dissociated by selective brain damage and which are commonly measured by intelligence tests. The first is *simultaneous* processing, which involves the synthesis of elements into groups and entails holistic, spatial representations. The second is *successive* processing, which, by contrast, takes place in a temporal sequence. These two modes of processing are thought to be, primarily, right- and left-hemisphere functions, and are sampled by different kinds of intelligence tests. Tests of simultaneous processing are thought to include Raven's Matrices (despite the belief of others that this is a pure *g* measure), memory for designs, and tests of visualization. Das *et al.* (1979) have noted that traditional tests of spatial ability are highly correlated with simultaneous processing. Tests of successive processing include digit span, serial recall, and many tests of verbal memory.

If simultaneous processes can be equated with spatial ability and successive processes with verbal ability, then the analysis of Das and colleagues is in broad agreement with the conclusion from psychometric data on the general nature of the major specific abilities.

(2)   There is also some evidence that *Performance* and *Verbal* IQ in the Wechsler test batteries can be dissociated in some cases of brain damage. Dennis (1985), in a study of 407 adult subjects who had early brain damage, found that the predictors of Performance and Verbal IQs differed. For example, diffuse brain damage was associated with a lower Performance IQ, but seemed not to affect Verbal IQ. Laterality of lesion was a good predictor only of Verbal IQ, but this was confined to a left temporal lobe locus.

The mapping between the simultaneous/successive distinction and Verbal/Performance IQ, although not isomorphic, is close enough to suggest some commonality. For example, the Performance scale of the WAIS (so called because the response mode is non-verbal) includes *Picture Completion*, *Block Design*, and *Object Assembly*, subtests which all involve, plausibly, some kind of visuo-spatial abilities. The Verbal scale is composed of *General Information*, *General Comprehension*, *Arithmetic*, *Similarities*, *Digit Recall*, and *Vocabulary*, subtests which involve, plausibly, verbal reasoning and sequencing processes. The differential influence of brain damage on Performance and Verbal IQ gives some tentative support for the proposition that verbal and spatial abilities constitute the major specific abilities. Indeed, the distinction between verbal and spatial abilities runs through most of the neuropsychological data on the differences between the two cerebral hemispheres.

(3)   The notion that the right and left cerebral hemispheres might support different kinds of intellectual functions has been a popular speculation (Ornstein 1972). The left-hemisphere is thought to be logical, verbal, and analytic, whereas the right hemisphere is intuitive, emotional, and holistic. Many techniques have been used to explore possible differences in cognitive style of the two hemispheres. The three major ones are: (a) experimental studies which show that some stimuli are processed more efficiently in one hemisphere than the other; (b) the influence of handedness on pattern of cognitive abilities; (c) the study of 'split brain' patients.

(a)   The basic assumption underlying the experimental work is that if the hemispheres process stimuli in different ways, then hemisphere functions can be inferred from the nature of the tasks that show an advantage for stimuli presented in the sensory fields that project to each hemisphere. This literature is too voluminous to review here (see Springer and Deutsch 1981, Milner 1971, Corballis 1986), but the major conclusion from this work, for current purposes, is straightforward: for the majority of people there is a right-hemisphere advantage for visuo-spatial memory and processing and a left-hemisphere advantage for verbal memory and processing.

(b)   Differences in the degree to which an individual prefers the left or right hand for certain tasks is taken by some to indicate differential development of the left or right hemisphere, and this, in turn, is hypothesized to have effects on cognitive abilities (Geschwind and Gallaburda 1987). In particular, since left-hemisphere functions are claimed to be primarily verbal, whereas right-hemisphere functions are primarily visuo-spatial, left-handers are hypothesized to show deficits in the former but superior ability in the latter. For example, Geschwind and Gallaburda claim that there is evidence to suggest that left-handers (who, *ex hypothesi*, have less developed left hemispheres) are more apt to report speech and reading problems than right-handers, and that there are more left-handers among architects (who, it is argued, require superior visuo-spatial skills) than we would expect, given the proportion of left-handers in the general population. However, the relationship between laterality of function as indicated by handedness and cognitive abilities is complex and contentious (Annett and Manning 1989, Bishop 1990). Suffice it to say that any abilities that are, arguably, discriminated by handedness seem to fall on one side of the, by now, familiar verbal/spatial divide.

(c)   Perhaps the most spectacular of all the studies of hemisphere differences are those of patients who have had their corpus callosum, the part of the brain that connects the two cerebral hemispheres, removed (usually in the attempt to control epileptic seizures). Levy and colleagues (1972) analysed experiments involving such patients, and argued that the two hemispheres seem to use different modes of processing. For example, if a range of objects presented to a single hemisphere (by keeping it in either the right or the left visual field) had to be matched with a picture, then patterns of errors differed between the hemispheres. If the objects were difficult to discriminate visually but had clearly associated verbal labels, then superior matching occurred when they were presented to the left hemisphere. By contrast, if the objects had similar verbal labels (in terms of their semantic relatedness) but were visually dissimilar, then matching was superior when the objects were presented to the right hemisphere. Since these heady early days, the 'split brain' story has become more problematic. (To what extent do cortical connections remain? Are the findings applicable to non-damaged populations?) Yet, these patients provide compelling evidence that, whether it be due to anatomical differences between hemispheres or something else, processing can occur in two distinct modes, roughly labelled as verbal and visuo-spatial.

The psychometric data had suggested that there are at least two specific abilities, usually labelled verbal and spatial, and it seems as if the same conclusion can be drawn from the neuropsychological literature. Although the conclusions regarding cognitive differences between the two hemispheres are dogged by difficulties in interpretation (Hissock and Kinsbourne 1987), a reasonable conclusion is that for the large majority of right-handers the processing of language seems to be a left-hemisphere function and that there is a tendency for visuo-spatial processing to be the province of the right hemisphere. There are, of course, other interpretations of these clinical findings, particularly concerning the role of verbal ability. It may be, for example, that the left hemi-sphere appears to be associated with verbal ability only because it instantiates the linguistic competence that I attributed to a *module* in chapter 4. Clearly, we need to be more confident that the processes underlying putative specific abilities in the domains of psychometry and neuropsychology map onto the same theoretical constructs. This is, of course, the goal of this part of the theory of the minimal cognitive architecture. But before a cognitive analysis of the nature of these underlying processes is offered, another example from a different disci-pline should convince even the most sceptical that there are reasonable grounds for supposing there to be at least two *specific processors* that underlie the ubiquitous distinction between verbal and spatial abilities.

Behaviour Genetics

The most researched influence of genes on cognitive abilities concerns the heritability of IQ. Currently, the best estimate of the contribution of genetic variation to variation in IQ is that it is about 50 per cent (Plomin and Daniels 1987), and much of this variation is likely to be due to the heritability of 'general' intelligence (Jensen 1987a). But I do not want to re-run the heritability of IQ debate here. Rather, I want to concentrate on the evidence in favour of genetic contributions to specific abilities.

In a review of genetic influences on learning disabilities and disorders of speech and language, Pennington and Smith (1983) marshal evi-dence from numerous studies in favour of specific genetic etiologies for dyslexia, including the different subtypes 'auditory verbal', 'visuo-spatial', and 'mixed'. Pennington and Smith (1983) point out that researchers vary as to how dyslexia is defined. For example, some use speech problems as a defining characteristic, while others specifically exclude them as diagnostic criteria. So the technique of identifying a specific disorder and *then* asking whether it has a genetic basis is unlikely to prove definitive in our search for specific abilities, depending as it does on an *a priori* analysis of what constitutes a specific ability. This kind of genetic analysis will always be limited by the appropriate-ness of the descriptions of behavioural symptomology for a cognitive conception of a specific ability.

The best evidence for specificity comes not from the search for a pos-sible genetic basis of some (perhaps contentious) specific ability, but from specific deficits seen in individuals of *known* genetic abnormality. This holds particularly if the nature of the genetic disorder can be con-trasted and mapped onto equivalent contrasts in cognitive abilities. Just such a situation is afforded by the data on the influence of sex-chromosome abnormalities on cognitive abilities.

The normal genotype has 46 chromosomes, two of which code for sex; one X and one Y chromosome (so called because of their shape) result in the normal male genotype (46, XY), and two X chromosomes result in a normal female genotype (46, XX). In groups in which there is a sex-chromosome abnormality, a specific cognitive deficit *always* results, although general intelligence is often affected as well. An extra sex chromosome (be it 47, XXX; 47, XXY; or 47, XYY) produces lower verbal than performance IQs and an increase in speech and language disorders. The most affected group are the 47, XXY boys. In fact, there seems to be a reciprocal relationship between the presence and absence of an X chromosome and the relative development of verbal and non-verbal (or spatial) abilities (Rovet and Netley 1983). So, for example, males with Klinefelter's syndrome, in which there is an extra X chromosome (47, XXY), have lower verbal than performance IQs (Netley and Rovet 1982). By contrast, females with Turner's syndrome, in which one X chromosome is missing (45, X0), seem to have a specific deficit in spatial ability. It seems likely that the X chromosome is implicated in the differentiation of verbal and spatial skills. Consistent with this is the pattern of deficits found in the recently discovered 'Fragile X' syndrome, which is the second largest known genetic cause of mental retardation after Down's syndrome.

Fragile X is an abnormality on the X chromosome, and its effects are therefore sex-linked; that is, they are expressed in males and carried in females. The main feature of affected people is mild to severe mental retardation. However, in a number of studies, Hay and colleagues have shown that not all abilities are equally affected. By and large, verbal skills are relatively unaffected but spatial tests show severe deficits, to such an extent that many Fragile X individuals would be classed as retarded on a block design test but not on vocabulary (Theobold *et al.* 1987).

A safe conclusion from the behaviour genetics literature is that there are abilities more specific than general cognitive ability. Consistent with our short reviews of both psychometric and neuropsychological data, the evidence suggests that verbal and spatial abilities can be differentiated by genetic etiologies.

Time Out

This review of the evidence from psychometric, neuropsychological, and behaviour genetics for cognitive abilities that are more specific than general intelligence was aimed at the first two of our questions concerning specific processors: namely, how many are there? and what is the nature of their computations?

Clearly the evidence so far converges on there being at least two specific processors, one that acquires knowledge using processes and representations that can be described, loosely, as verbal and another that acquires knowledge using processes and representations that can be described as visuo-spatial. We shall see in a moment that the evidence from cognitive studies of different modes of processing maps onto this very distinction.

The idea that there are at least two specific abilities may seem uncontroversial. But why restrict the number to two? For example, a case could be made from one, perhaps even two, of these research areas for considering 'reading' or 'number' as a specific ability. However, postulating a specific processor for every specific ability suggested by any research area is exactly what I want to avoid doing – and with good reason.

1   The point of using converging evidence from these diverse disciplines is to aid in the specification of the *minimal* cognitive architecture. Our strategy must, therefore, be a conservative one. Only specific abilities that are common to *all* these diverse disciplines are candidates for having a basis in the operation of a specific processor. For example, psychometricians may well argue that factor analysis has clearly shown that problems requiring verbal comprehension are clearly distinct from problems requiring some other verbal abilities; say, solving verbal analogies. But if this distinction has no basis in neuropsychological evidence or behaviour genetics, then it is likely to be one that is not worth pursuing. In any case (as we shall see in a moment) we cannot know from the psychometric data alone whether the psychometric distinction is based on a different cognitive mechanism; and only different *cognitive* mechanisms are of any interest. The technique of requiring converging evidence minimizes the number of specific abilities that must be accommodated by the theory and prevents it from becoming a facile hotchpotch of whatever anyone, anywhere, at anytime has considered to constitute a specific ability.

2   As we saw in chapter 4, many so-called specific abilities may be underlain by a *module*. However, remember that a module is an all or nothing device that shows no normally distributed individual variation. We are looking for reliable, common examples of particular types of specific abilities: those that show normally distributed variation in the general population. The evidence is robust for only two such abilities.

Many other specific abilities/deficits ('reading' being, perhaps, the best example) may be based on modularized processes rather than specific processors (see chapter 8).

3   A specific processor is a knowledge-acquisition mechanism. This is a very powerful construct. I do not want to be in the position of positing a different mechanism for each specific piece of knowledge. Powerful mechanisms can easily become weak explanatory devices. There are many ways in which specific differences in *knowledge* (which under some descriptive frameworks constitutes an ability) can arise other than through different kinds of processing mechanism underlying them. The most obvious source of variation, of course, is the kind of environment to which individuals are exposed. Maybe 'Number' is a Primary Mental Ability in the sense used by psychometricians, not because there is a specific processor for numerical calculations but because there are specific and systematic differences in the ways in which individuals are exposed to the kinds of information on which numerical knowledge depends.[2] We are searching for underlying knowledge-acquisition *mechanisms* which have characteristically different operating principles. The same mechanism could be responsible for many different kinds of specific *knowledge*; therefore, specificity of knowledge may be an irrelevant clue for a specific processor.

Using converging evidence that crosses different levels of explanation (as in this chapter), which refers to psychometric, neuropsychological, and genetic data, is in many ways merely preliminary to the main job. It simply provides us with a fast way of finding the right ball park. To the extent that we want to put our money on the proposition that specific abilities are underlain by different kinds of processing mechanisms, then the evidence from psychometry, neuropsychology, and behaviour genetics suggests that the verbal/spatial distinction is the safest bet. We must also be aware of the dangers, discussed in the previous chapter, of using evidence from different research disciplines where we cannot be sure whether the constructs (kinds of specific abilities) map onto each other. Do the specific abilities suggested by differences between Verbal and Performance IQ, between verbal and spatial ability, between left and right hemispheres, and between different genetic causes map onto the same *thing*? Because I have argued that only a cognitive theory can accommodate the agenda for a psychological theory of intelligence, it follows that these constructs merely provide a scaffolding for theory

---

[2] In this context we should also consider other non-cognitive influences on knowledge acquisition, such as personal attributes like motivation and personality.

building. The theory itself will stand or fall, ultimately, on the strength of the cognitive evidence. It is to this evidence that I now turn.

Cognitive Processes

Cognitive psychologists are most at ease with the class of thought processes that might constitute a *verbal* dimension of individual differences; that is, thought processes that are sequential and, in all probability, propositional. The classic process of this kind would be language itself; but most reasoning and inferential processes (deductive, inductive, analogical, syllogistic) fit this category of thinking. Indeed, for some, what distinguishes *thought* from other psychological processes is its language-like structure: to *think* is to process symbols in sequences which have a syntactic structure (Fodor 1975, Pylyshyn 1984). Perhaps it is because many of our thoughts seem to constitute a kind of inner speech that this mode of processing seems ubiquitous and natural. Certainly it would be superfluous to provide documented evidence for the existence of such processing. Rather, it is the possibility that there could be *any other* kind of process that has exercised cognitive psychologists, particularly over the past fifteen years or so. What might another kind of process be like? Given the discussion in this chapter so far, you may well have anticipated one answer. Instead of thought always being sequential and propositional, might it not sometimes be holistic and *imaginal*? In fact, is this not the distinctive characteristic of visuo-spatial *imagery*? And is imagery not as much a *thinking* process as is, say, deductive inference?

There are, of course, many historical antecedents of the belief that visuo-spatial imagery is a distinctive thought process. Unfortunately, these antecedents are bound up with ferocious disputes over the kinds of phenomena that psychology should try to explain while remaining a strictly scientific discipline. To the experimentalists of the nineteenth century, not only was imagery considered a *bona fide* thought process, but, in many ways, it was regarded as the very stuff of psychology. The method of introspection was based on the premise that the contents of experience are illuminating for theories of mental processes. For pioneers such as Titchener it was inconceivable that psychology would not have the conscious experience of imagination as a central explicandum. But others feared that a concern with *imaginal* processes might lead to flirting with illusion.[3] Those fears were most firmly expressed by the behaviourist school of J. B. Watson. For thirty years the hangover from

[3] It is said that when Galton asked some of his scientific colleagues whether they had mental images, he had great difficulty in communicating what he could possibly mean,

the heady days of introspection was a denial of the scientific respectability of all notions pertaining to any kind of mental state, let alone processes of imagery. The effects of this hangover can still be detected in current disputes over the status of mental images for cognitive theories of mental processes. Are the processes peculiar to imagery fundamentally different? Or are they simply seductive epiphenomena of some standard propositional ones? We will address this most fundamental issue in a moment.

It is fitting, then, that of the many forces that led to the breakup of the behaviourist stranglehold on psychology and the subsequent rise of cognitive psychology, it was the re-emergence of the explanatory power of images *within* the behaviourist school that most obviously signalled the need for a mentalist explanation. Thus, the earliest precursors of truly cognitive theories were those theories of rat maze learning which proposed that behaviour was controlled on the basis of cognitive maps (Tolman 1933). Those cognitive maps are easily recognized today as spatial *representations*. Somehow the idea that behaviour might be controlled by propositional 'representations' would not have seemed as heretical to the behaviourist hegemony as was the idea that some kind of image controlled behaviour.[4]

It has become clear over the past thirty years, as information processing psychology has developed into cognitive science, that the central theoretical issue is the nature of mental representations. Paivio (1971) proposed that mental representations come in two forms: verbal and visual. He considered that words and pictures could be represented in either code, and that it may be easier to remember words that have obvious associated images (so-called concrete words) than those that do not (so-called abstract words). Similarly, it should be easier to remember pictures that have obvious verbal labels. Many empirical justifications of this view were amassed. However, while not denying that the experience of images is real enough, other cognitive theorists disputed that verbal and visuo-spatial *representations*, or indeed processes, are qualitatively different and denied that *images per se* could play a *causal* role in thought (Pylyshyn 1973, Clark and Chase 1972). What can we conclude about the distinctiveness of visuo-spatial processing on the current evidence?

---

given that all the illustrious gentlemen claimed no experience of such images.

[4] Although control of behaviour by *any kind* of knowledge system is, of course, just as heretical. However, many behaviourists tried to save stimulus–response psychology with constructs such as internal stimuli and responses that, in a different age, have striking similarities with the condition–action pairs of production systems, an early flagship of one branch of cognitive science.

As always, the story is more complicated than might at first appear. In particular, it may be empirically indeterminate whether what distinguishes these processes from the more accepted propositional/verbal ones is the nature of the mental representations underlying them or the way in which these representations are processed (J. R. Anderson 1978). Certainly, one of the most comprehensive current accounts of imagery processes proposes that there is a considerable overlap between verbal/propositional and visuo-spatial processing (Kosslyn 1981, 1983, 1988). Kosslyn emphasizes that imagery can no longer be considered as a single process. He argues, for example, that visuo-spatial representations out of which 'images' must be constructed come in two clearly different forms. The two forms are determined by different functional properties of images. The first is the 'what' property: the kind of spatial information that is useful for shape analysis and object recognition. Some experiments have shown that if shapes are constructed and compared in the imagination, then the imagined shapes are themselves constructed sequentially out of stored constituent parts, in a manner akin to a standard sequential propositional process. On the other hand, 'where' information – that is, the representation of location within a frame of reference – is more clearly visuo-spatial. If we are asked whether Stoke is approximately on a straight line between Edinburgh and Bristol, it is probably easier to imagine a map of Britain and trace a line between Edinburgh and Bristol and see if Stoke comes anywhere near it, than to calculate the truth value of this proposition from stored information about the distances between towns.[5] For such a task it is obviously advantageous to have the information represented directly and simultaneously in an image.

So imagery subsumes many different kinds of processes and representations involved in storing, generating, and interrogating visuo-spatial images, many of which are common to the more traditional verbal/propositional thought processes. However, Kosslyn argues that what is distinctive about images is that they can *depict*, whereas propositions can only *describe*. Propositions are *abstract*, with *discrete symbols* that have combination rules determined by a *syntax*. Images, on the other hand, are *concrete* (we can imagine only a particular exemplar from an abstract category like 'dog', not the abstraction itself), with no symbols or syntax. Kosslyn claims that images can be shown to share properties of depictions; principally, that many physical properties of

[5] Even though the algorithm is simple. If the distance between Edinburgh and Bristol approximates the sum of the distances between Edinburgh and Stoke and Stoke and Bristol, then Stoke must lie on a straight line between Edinburgh and Bristol (taken from example given in J. R. Anderson 1978).

visuo-spatial characteristics of objects are depicted in their imaged representation. For example, scanning time for information retrieved from an image is directly related to the size and other spatial qualities of the objects being imaged. This would not be expected if the image were some kind of epiphenomenon of a process which operates on an underlying propositional representation.

Since the major controversy over whether visuo-spatial thought is characteristically different from propositional thought centres on their respective representational formats, it is interesting to note that the most unambiguous evidence for distinctiveness comes from theories not of how images are stored and constructed, but of how they are transformed, and in particular, of how they are rotated.[6] If pictures of two objects, side by side, are presented to subjects and one is a rotated version of the other, the time it takes subjects to affirm that the objects in the pictures are the same is a linear function of the degree of displacement of the orientation of the second object from the first (Shepard and Metzler 1971, Shepard and Cooper 1982). The clear suggestion from these studies is that the image of the second object is being 'rotated' until it appears at the same orientation as the first, whereupon it is compared for identity. If this is an accurate characterization, then, at least at some level of analysis, such a process must be uniquely visuo-spatial; it is hard to imagine how or why we might want to rotate a proposition, or what it would 'look like' if we could! It is interesting, then, that this process is the one that has been found to relate most consistently to psychometric measures of spatial ability. The best way to evaluate this evidence and to give some idea of the complexities inherent in making connections between cognitive descriptions of visuo-spatial abilities and psychometric ones is to give a more detailed account of a study which attempts to link the two.

Just and Carpenter (1985) attempted to account for differences between low and high scorers on a test of psychometric spatial ability in terms of an information-processing model of mental rotation. To do this, they investigated the details of subjects' performance on specially constructed exemplars of items from the Cube Comparison test (a test from Thurstone's Primary Mental Abilities battery). The stimuli consisted of pictures of two cubes, side by side, with letters on each face (see figure 5.1). The task was to decide if the two cubes, each with only three of its faces visible, could be pictures of the same cube. The difference between the cubes was varied on three dimensions: (1) the number

---

[6] Although such rotations, which appear to be obviously simultaneous and analogue, can always be explained by a suitably ingenious propositional mechanism (Pylyshyn 1973, 1979).

of matching faces – that is, faces containing the same letters; (2) the degree of rotation (0°, 90°, or 180°) necessary to re-orient the matching stimulus to the standard; (3) whether the rotation could be performed only on a standard trajectory – that is, around an axis perpendicular to one of the faces – or whether there was an alternative rotation, shorter than the standard, around an axis not perpendicular to one of the faces.

By examining reaction times and protocols of subjects' reported strategies, Just and Carpenter constructed an information-processing model of performance on this task. This model emphasizes that 'visual imagery' subsumes a number of distinct cognitive processes. Since the Cube Comparison test is known to be a good measure of what psychometricians mean by spatial ability, their analysis provides us with the opportunity to map a psychometric conception of 'spatial ability' onto a cognitive one. The result is interesting.

First, there are clearly a number of different *strategies* that can be employed to solve such problems. In particular, subjects differed in the cognitive coordinate system that they used for comparing the cubes. For example, one high-spatial ability subject reported using what Just and Carpenter call an 'orientation-free' coordinate system. This involves encoding the relations between letters on each cube exhaustively (for example in figure 5.1(a), the top of K points to the bottom of M, the right side of M points to Q, and so on), thereby obtaining a description of the structure of each cube that is independent of its orientation. In effect, this removes the need to rotate either cube, and may make performance more heavily dependent on propositional thought processes than on distinctively visuo-spatial ones. So, paradoxically, *some* subjects may be high on psychometric spatial ability due to the adoption of verbal/propositional performance strategies.

Second, there was no difference between high- and low-spatial ability individuals on a stage of processing identified with a search process – that is, a process whereby potentially identical faces are located on each cube. Thus, not all processes concerned with scanning and matching visuo-spatial displays are related to psychometric spatial ability.

Third, the major differences between high- and low-spatial ability subjects are in the processes Just and Carpenter call 'rotation' and 'confirmation'. Once potentially matching faces are located on each cube (the search process), a selected face must be rotated to match its target face on the reference cube. Subsequently, the two other faces must be rotated along this same trajectory to check for a match with the reference cube. It is the efficiency of this process that most distinguishes subjects who score well from those who score poorly on psychometric tests of 'spatial' ability. Just and Carpenter estimate that

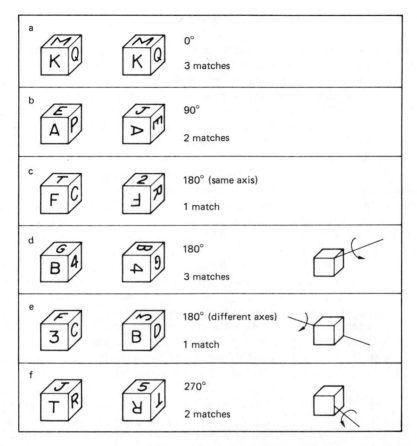

Figure 5.1   Stimuli used by Just and Carpenter 1985. Copyright 1985 by the American Psychological Association. Adapted by permission.

high-spatial ability subjects can rotate 'images' at twice the rate of low-spatial ability subjects.[7]

Conclusion from Cognitive Processes

Although there are many intricacies and qualifications, it seems reasonable to suppose that there is something different, in terms of the kinds of cognitive mechanisms they employ, between visuo-spatial and

[7] The sting in the tail of their analysis, however, is that their own simulation of the rotation process is propositional, and uses their general-purpose CAPS Production System model.

verbal/propositional processing. It is surely more than a coincidence that the same distinction has carried us all the way through the chapter. For this reason I feel safe in concluding that there are two specific processors, one of which is more implicated in propositional-like thought, while the other is more implicated in visuo-spatial-like thought. Note that I do not want to claim that one is a verbal processor and one is a spatial processor. At this stage of development of the theory, I can do little more than speculate as to what processing dimension (whether it be the nature of their operations or the nature of the representations over which they operate) distinguishes their operation. It is better that the two specific processors be called simply SP1 and SP2, with SP1 more implicated in verbal/propositional processing and SP2 more implicated in visuo-spatial processing. I do not say this out of pedantry, for it is not simply a terminological issue, but one of considerable theoretical importance, which I shall develop later. This is reinforced by the other major conclusion that we can take from the review of the relevant literature on cognitive processes.

It is clear that the correspondence between the cognitive processes considered to distinguish visuo-spatial from verbal/propositional processes and psychometric conceptions of what constitutes spatial ability is not unproblematic. It may be that an individual has high-spatial ability, as determined by a score on a psychometric test, because the distinctive mechanisms underlying visuo-spatial processing are particularly efficient: for this individual SP2 would be particularly powerful. But a high-spatial test scorer could also be so because SP1, the specific processor underlying verbal/propositional thought, is very powerful, whereas SP2 is actually rather weak! In this case, because the 'visuo-spatial processor' is so poor, the individual adopts a processing strategy based on a propositional process (like the orientation-free description) generated by SP1. Such a strategic choice is, of course, possible only if problems are solved by mechanisms that have universal processing power (Turing machines), exactly the property attributed to specific processors at the beginning of this chapter.

The major lesson is this: it is dangerous to label SP1 and SP2 as verbal and spatial processors, because these terms are not as yet computationally distinguished, so we do not gain anything in theoretical specificity and we may lose something as a result. What we may lose is the clear distinction between these constructs as hypothesized mechanisms in a cognitive architecture and the non-isomorphic psychometric constructs of verbal and spatial ability. Simply put, we run the risk of confusing two things: a *computational* account of ability (what the important kinds of *information- processing mechanisms* that display individual differences are) and a *psychometric* account of ability (what the major dimensions of individual differences in intelligence test performance are). And as we have seen, the cognitive and the psychometric construct of an ability are certainly not the same thing.

The Final Component

The final component of the cognitive architecture is now in place (see figure 5.2). The goal of this chapter was to specify the nature of the knowledge-acquisition mechanisms that are constrained by the speed of the *basic processing mechanism*. I argued that these mechanisms were unlike *modules* in two ways: (1) they contribute to individual variation in cognitive abilities; (2) they deal with idiosyncratic knowledge-acquisition and thus are like Turing machines or general purpose problem-solvers. Having called these mechanisms *specific processors*, other questions followed: How many specific processors are there? What is the nature of their computations? And how are they constrained by the speed of processing of the basic processing mechanism?

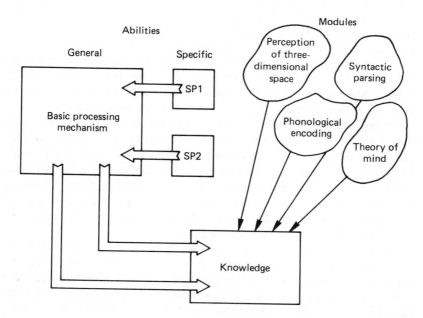

Figure 5.2 The specific processors (SP1 and SP2) underlie individual differences in specific abilities, and complete the minimal cognitive architecture that can account for individual differences in intelligence.

The first question has been unequivocally answered: there are two specific processors.

As for the second, all we can conclude is that each processor acquires knowledge in a characteristically distinct way. One is more implicated in the acquisition of the kind of knowledge that has been described as verbal/propositional, whereas the other is more implicated in the acquisition of visuo-spatial knowledge. Any attempt at a proper account of the nature of the specific processors (that is, a complete computational description of how they operate) is obviously premature, since this would be equivalent to proposing a full-blown computational theory of both verbal and spatial processing, which is well beyond the scope of this book. However, it might be helpful to think of the specific processors as having the functional characteristics of computer languages. For example, we could adopt the hypothesis that SP1 was a bit like LISP, a powerful propositional list processor, whereas SP2 was more like FORTRAN, which, while still a propositional language, is not as good as LISP at list processing, but is much better for engineering calculations. Computer languages have the admirable properties of Turing machines; that is, any problem that can be solved with a LISP program can also be solved by FORTRAN, and vice versa, it is just that *one language is more suited to a certain class of problems than another*. Thus, a weak version of LISP would probably solve most list processing problems more easily than would FORTRAN, no matter how powerful the version of FORTRAN. Nevertheless, all other things being equal, FORTRAN would get there in the end. However, as we shall see in a moment, for the specific processors, not all other things are equal.

Although this analogy should not be taken too seriously, it does capture some of the essential features of the specific processors: namely, that they have universal computing power and are *not* verbal and spatial processors, but rather problem-solving mechanisms whose computational architecture is more suited to one or other of these classes of knowledge. Finally, the analogy gives us a way of approaching the $64,000 question: How are the specific processors constrained by the speed of the basic processing mechanism?

There are, potentially, a number of kinds of answers to this question. The best would be a mechanistic account which says something like: 'Because the computations of SP1 and SP2 involve such and such and because the speed of the basic processing mechanism affects this and that, this means that the computations of SP1 and SP2 will be less constrained at higher processing speeds.' Specification of 'such and such' and 'this and that' is *the* answer to the $64,000 question. It is no surprise that we are nowhere near answering it; for (a) we have only just asked the question, and (b) it requires a full computational theory of the operation of SP1 and SP2. However, there are more tractable questions that we can begin to answer now. They concern the relationship between processing speed and the operation of the specific processors and how specific processors contribute to individual differences in intelligence. We know now, *ex hypothesi*, that the speed of the basic processing mechanism is the primary basis of individual differences in intelligence, but we also know that variations in the power of the specific processors contribute to individual differences too. Can we quantify their contribution? What is the effect of increasing processing speed on this contribution? What are the implications of this relationship for *patterns* of cognitive abilities? These questions will be explored fully in chapter 8 but for now let me give a short preview.

Hypotheses Concerning General Intelligence and Specific Abilities

The central hypothesis of this book is that when we *think*, we are implementing a problem-solving algorithm generated by a specific processor. It is this implementation that is constrained by the speed of the basic processing mechanism. So, more precisely,

> the observed correlation between the power of the SPs that we see in the phenomenon of general intelligence is caused by the constraint that the speed of the basic processing mechanism imposes on the implementation of the knowledge-acquisition routines generated by the SPs, such that at low speed, or efficiency, only the simplest routines can be implemented, irrespective of possible differences in the latent power of the SPs.

Individual differences in general intelligence are, then, the consequence of the constraint that the speed of the basic processing mechanism places on knowledge-acquisition routines generated by the specific processors. Individual differences in specific abilities, as manifested in knowledge or problem-solving abilities, will thus be a function of two *different* things:

1   the latent 'ability' (computational power) of a specific processor, which determines the availability of complex knowledge-acquisition algorithms,
2   the speed of the basic processing mechanism: the higher the speed, the more complex the algorithm that can be implemented.

Such a relationship predicts that specific abilities (the manifest expression of the latent power of the specific processors) will become increasingly apparent with increasing levels of general intelligence (speed of the basic processing mechanism). Consequently, at lower levels of intelligence, individual differences will depend more on variations in the speed of the basic processing mechanism. At higher levels of intelligence, where processing speed is faster, differences in the latent power of the specific processors will be more fully expressed (see figure 5.3), resulting in a greater proportion of the variation in intelligence being attributable to specific abilities.

This *differentiation hypothesis* will be dealt with more fully in chapter 8 when we consider in more detail the implication of the overall theory for the nature of mental retardation and specific cognitive deficits and talents. But for the moment, here are some of the principal reasons for believing that it may be true.

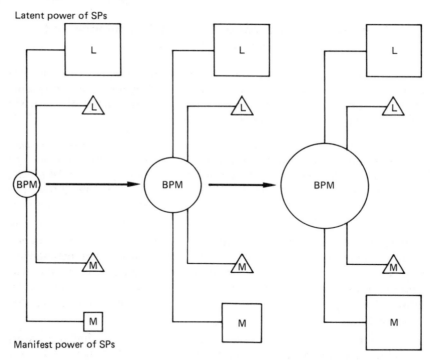

Figure 5.3 The differentiation hypothesis (L = latent power of a specific processor; M = manifest power of a specific processor; BPM = basic processing mechanism). As speed of the basic processing mechanism increases, the high latent power of one of the specific processors becomes increasingly manifest. Thus a difference in the latent power of the two specific processors may be obscured at low processing, whereas increases in speed of processing will lead to increasing differentiation of specific abilities.

1   Spitz (1982) has shown that mentally retarded people matched for mental age with normal control groups are *most* strongly handicapped on *g*-loaded tests and that *g* accounts for more of the variation at lower IQs.

2   Although *idiots savants* present a glaring exception (to be discussed in chapter 8), the general rule is that specific talents are most obvious at higher IQs (O'Connor and Hermelin 1981).

3   The relationship between low-level information-processing speed measures and IQ is greater at below-average IQ than at above-average IQ (Detterman and Daniel 1989).

There is, then, a *prima facie* case for supposing that the structure of abilities changes with level of IQ and that this phenomenon may be underpinned by the changing constraint that different speeds of processing impose on the specific processors. We will return to these issues in chapter 8.

The Firm of Anderson and Anderson

The story has become increasingly abstract over the last two chapters, and it may be that a metaphor which presents the sense of the theory from another angle will aid comprehension. On the other hand, many people do not like metaphors, finding them distractions rather than aids to insight. If you are that way inclined *or* you feel you understand the theory quite well enough already (*or*, indeed, if you are inclined to take metaphors too seriously), then you would be advised to skip to the next section.

Let us suppose that we are studying individual differences in law firms. Firms differ in success, and this is measured in terms of the briefs and contracts they produce. Some firms are clearly more successful than others. Our remit is to discover why this is so.

Our first, and most striking, observation is that some firms are *generally* better than others. If they tend to be successful in property conveyancing, they are also successful in divorce settlements and criminal cases too: the idea of the specialist firm turns out to be a bit of a myth (although there are always one or two exceptions). After considerable analysis of our data, we come to three conclusions:

(1)   The single most important contribution to individual differences in success of law firms is the speed of the firm's *secretary* (*basic processing mechanism*). The faster she can type, the more productive (and successful) the firm. This, of course, comes as a great surprise to the Law

Society (cognitive psychologists), whose members have long believed that what determines success is largely the *partners*, these being lawyers whose abilities depend, for the most part, on where they were trained and what their experience has been. The partners do have an important role, although, in terms of variation in performance accounted for, a lesser one.

(2)   But even the fastest secretary is of little use with nothing to do! It turns out that although many people work for the firm, the ones who have the skills necessary to generate new briefs and contracts are, indeed, the *partners* (*specific processors*). Typically, the partners bring complementary skills to the firm. For example, in the firm of Anderson and Anderson, the senior partner is rather conservative, but is very thorough and handles most of the highly technical work. The junior partner is rather adventurous, is not particularly good at dealing with the details of individual contracts, but has 'flair' and tends to be given the more unusual cases. So, the partners handle different aspects of the firm's work. However, both can, when necessary, handle each other's portfolio. In fact, in some firms one of the partners is so weak that the other partner carries most of the work-load. There again, in other firms

Figure 5.4 The secretary.

Figure 5.5 The partners.

the secretary is so slow that only very basic and simple cases are ever completed, irrespective of how talented one or both of the partners may be. A slow secretary can hide the fact that the partners differ widely in their abilities; it is only in the better firms that a partner's weakness becomes obvious.

(3)   Of course, to qualify for registration, every firm must provide standard but specialized services for their clients. For example, they must be able to represent a client in court. To do this, firms engage a *specialist* (*module*) who earns his or her living just by presenting cases in court.

Indeed, over the long history of the law, a whole body of specialists have evolved to meet these specialized needs. All firms have access to these specialists, whose services are, of course, uninfluenced by the firm's secretary, because they provide the firm with a finished 'product'. You might wonder why one of the partners does not do this work for the firm. It is simply not feasible for one of the partners to take on this kind of work, because he or she does not have the specialized knowledge necessary, and in any case the firm's secretary would not be able to keep up with the work-load involved: nothing else would ever get done!

Because all firms have access to specialists, they do not contribute to differences between firms, although they are vital for their operation. Very occasionally we find a firm that has not been able to contract the services of one of these specialists; but either the firm is just starting out and will soon employ one (as we shall see in the next chapter) or it is severely handicapped and struggling to cope.

The Minimal Cognitive Architecture

The minimal cognitive architecture necessary to accommodate the phenomena that are central to a psychological theory of intelligence is now complete. Three kinds of information-processing mechanism are involved in the acquisition of knowledge.

1   The *basic processing mechanism* varies in its speed, differentially constraining problem-solving algorithms generated by the *specific processors*. These two mechanisms are responsible for *thinking*, the process by which most knowledge is acquired. Because the basic processing mechanism varies in speed among individuals in the population, the constraint it imposes will generate the phenomenon of general intelligence. The relationship between the basic processing mechanism, the specific processors, and idiosyncratic conditions of information input explains (a) why some knowledge-free measures of processing predict knowledge-rich performance and (b) why individual differences in intelligence are general, but knowledge itself is domain-specific.

2   There are two *specific processors*, each of which acquires knowledge in a characteristically different fashion. They are akin to computer languages in that (a) they are like Turing machines, which means that they have universal computing power and are thus ideal for acquiring idiosyncratic knowledge, and (b) each one is suited to a different class of problems. The kinds of abilities they underlie are best classified as verbal/propositional and visuo-spatial. As the speed of the basic processing mechanism increases, and hence its constraint on the specific

Figure 5.6 The firm of Anderson and Anderson.

processors decreases, differences between these kinds of abilities will become more apparent.

3 Modules are necessary to explain why it is that not all knowledge is so constrained by the speed of the basic processing mechanism. Indeed, it is remarkable that some of the more complex computational problems discovered by cognitive science are the very problems that are unrelated to individual differences in intelligence. Modules are specialized, computationally complex mechanisms whose function is to provide us with evolutionarily prescribed information which could not be furnished by general-purpose problem-solvers – that is, by the specific processors.

This architecture accommodates three of the five crucial items on our agenda for understanding the nature of intelligence and development, listed in chapter 1. It explains why intelligence is general (item 3). Simultaneously, it accommodates the fact that there are differences in abilities more specific than general intelligence (item 4). Finally, it explains why there are some obvious exceptions to the rule that IQ predicts cognitive performance (item 5). Two major items remain: (1) why cognitive abilities increase with development, and (2) why it is that individual differences in intelligence are stable despite massive developmental changes in knowledge. To address these items requires the finishing touch to the cognitive architecture: it must be given a developmental dimension.

6

The Development of Intelligence?

The three kinds of processing mechanisms that constitute the minimal cognitive architecture that can account for individual differences in intelligence, the *basic processing mechanism*, the *specific processors*, and the *modules*, are shown in figure 6.1. A look at the diagram should make it clear that knowledge is acquired via two routes.

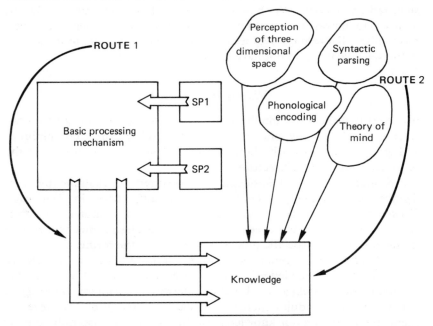

Figure 6.1 *The developmental routes. Route 1 knowledge is acquired by thinking. Route 2 knowledge is given directly by modules which mature at different times in development.*

Route 1 (left-hand side). What I called, in chapter 4, 'idiosyncratic knowledge' is acquired by implementing knowledge-acquisition routines generated by the specific processors. The speed of the basic processing mechanism constrains the complexity of the routines that can be implemented. Because the constraint of the basic processing mechanism is less at higher processing speed, differences between the latent powers of the specific processors will be more obvious at higher processing speeds and will account, therefore, for an increasing proportion of the variation in intellectual performance. When we acquire knowledge via route 1 we can be said to be *thinking*. Paradoxically, perhaps, idiosyncratic knowledge might also be called 'encyclopaedic knowledge', because much of it will be held in common by members of the same culture. Still, no two individuals will have the same encyclopaedic knowledge. Variations will be due, primarily, to individual differences in route 1 knowledge-acquisition mechanisms and also to differences in experience.

Route 2 (right-hand side). Knowledge that is evolutionarily invariant is acquired by *modules*. As shown in figure 6.1, the modules carry out their computations unconstrained by the speed of the basic processing mechanism. Knowledge acquired through route 2 is given to us *directly*, in the sense that we do not have to think to acquire it.

What is Intelligence?

It is clear that the term 'intelligence' as it is used in psychology does not refer uniquely to any one aspect of this architecture. To minimize ambiguity, I should clarify how two of the more common uses of the term map onto the architecture.

Individual differences in intelligence, as measured, for example, by intelligence tests, actually reflect individual differences in *knowledge*. However, what I mean by 'intelligence' in regard to individual differences is variation in the speed of the basic processing mechanism and in the power of the specific processors (primarily the former).

I will use the more generic sense of the term 'intelligence' to refer to the contents of knowledge itself. In particular, when I refer to the development of intelligence, I will be referring to changing *knowledge*. I will argue that modules increase intelligence *directly* during development (by increasing our knowledge of the world) and also indirectly by affording new *modes of representation*.

The distinction between 'intelligence' as a property of thinking (route 1) and 'intelligence' as a reflection of the contents and structure of knowledge (which depend in part, of course, on thinking) is crucial. It is crucial not only because modules contribute to intelligence *qua* knowledge, and not to intelligence *qua* thinking, but also because it emphasizes that intelligence (*qua* knowledge) can develop without the need to postulate any change in the mechanisms underlying intelligence (*qua* thinking). Indeed, in this chapter I will argue that modular change is fundamental to cognitive development and that it is one of the principal ways in which cognitive development can be distinguished from knowledge-acquisition. But I am running ahead of myself.

Explaining the Agenda

The minimal cognitive architecture allows us to resolve some of the conundrums surrounding the concept of intelligence mainly because it draws attention to the fact that psychometric 'intelligence' is not synonymous with knowledge. Intelligence tests work, by and large, by measuring variations in the quality and quantity of the knowledge base. However, the primary cause of these differences is the speed of the basic processing mechanism. This is why 'intelligence' is general, but knowledge itself is domain-specific. It also explains why 'intelligence' can be measured by relatively *knowledge-free* information-processing tasks. Moreover, and in contrast to other theories that have general intelligence as a central construct, this architecture explains why all knowledge is not related to measures such as IQ. Knowledge that has an invariant data base and substantial cash value in our evolutionary history is provided by specialized processing modules. Since the computations of modules are unconstrained by processing speed, it resolves the conundrum that it is the most *computationally* complex psychological functions (for example, visual perception and language acquisition) that show the most independence from IQ, while it is the simplest computational functions that are heavily represented on tests of intelligence. Similarly, damage to a module can explain why some cognitive deficits, such as dyslexia, are found in individuals of high IQ.

Figure 6.1 represents the minimal cognitive architecture that can account for what we know about individual differences in intelligence and knowledge. The remaining items on the agenda laid out in the first chapter concern development. Why does development appear so homogeneous across many different knowledge domains? Why are individual differences so stable despite large absolute changes in the quantity and complexity of knowledge available over the developmental period? It is the goal of this chapter to specify the developmental dimension of the theory.

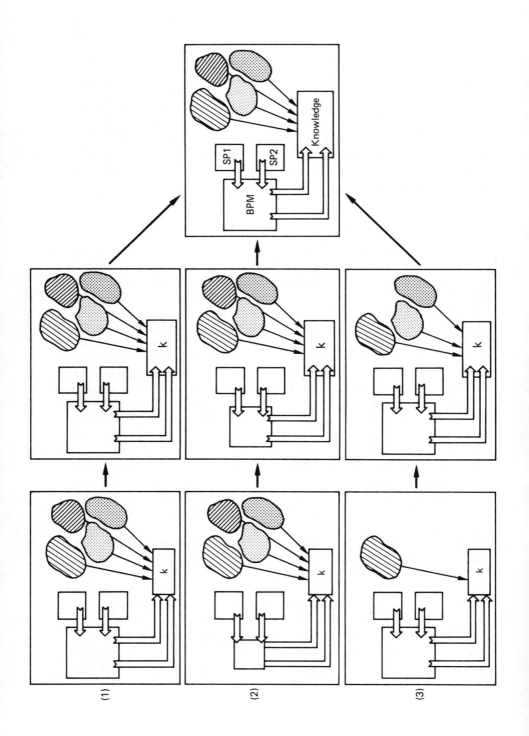

(1)

(2)

(3)

Figure 6.2 Developmental options. It may be that nothing changes in the architecture (option 1). Given that the mechanisms are designed to acquire knowledge, 'development' may, simply, be the name we give to this process. In this case, development would be a gradual and continuous process of knowledge acquisition, albeit one which is constrained by the speed of the basic processing mechanism, thus intimately related to individual differences in intelligence. In fact, this option makes no principled distinction between knowledge acquisition and development or between the psychometric constructs of IQ and mental age (MA), and is clearly antithetical to the kind of cognitive development typified by Piagetian theory. However, it may be that the speed of the basic processing mechanism increases during development (option 2). Since I have argued that this mechanism imposes the major constraint on thinking, not only will development reflect a gradual accumulation of knowledge; it will also reflect a decreasing constraint on the complexity of thought. To the extent that the increase in complexity of thought is stage-like, this option may be compatible with a qualitative-change view of development. But developmental level and individual differences would then be based on the same underlying factor: variations in speed of processing. The third possibility is that new modules may be acquired during development (option 3). This would allow radically new kinds of knowledge to be acquired and would predict abrupt discontinuities in underlying competence. Further, since modules are unconstrained by the speed, or efficiency, of the basic processing mechanism, development would be independent of individual differences in intelligence.

A Specification of the Developmental Dimension

The theory as it currently stands offers a number of developmental options (see figure 6.2). It may be that there is nothing special about development: that is, that because the mechanisms already specified are *knowledge-acquisition* mechanisms, changing knowledge is an inherent property of the architecture. On the other hand, we might suppose that one, two, or all three of the mechanisms might be subject to developmental change. A number of possible questions follow from such speculations. What would the consequences be if speed of processing changed during development? Do the specific processors increase their computational power? Do new modules come on-line at different times during development? Clearly, the theory provides powerful developmental options. However, there are considerable *a priori* constraints on the specification of the developmental dimension:

1   The most obvious is a data constraint. This is a more powerful constraint than is usual, in that the theory must account for data that are two-, rather than one-dimensional. The theory must not only account for data on cognitive development but must also account for the *relationship* between cognitive development and individual differences in intelligence. As we shall see, having to satisfy, simultaneously, the data from both dimensions reduces the available options.

2   The second constraint is a theoretical one: the developmental dimension must preserve the coherence of the theory as it applies to individual differences. The developmental dimension cannot violate an axiom of the individual differences model in order to satisfy developmental data. For example, if a developmental change is to be explained by the coming on-line of a new module, this developmental change should be uninfluenced by individual differences in intelligence, because modules are unconstrained by variations in the speed of the basic processing mechanism. We shall see that this proves to be a powerful constraint.

There are polar positions that can be taken regarding the relationship between individual differences in intelligence and cognitive development. One is that, whether intelligence is best regarded as a low-level property of biology or a high-level property of cognition (see chapter 3), there is no principled distinction to be made between individual differences in intelligence and the development of intelligence. The other is that, on the contrary, the development of intelligence and individual differences in intelligence are independent constructs, there being no connections between the two versions of intelligence either theoretically or empirically. We will examine each position in turn.

Polar Position 1. Whatever process underlies individual differences also causes developmental change. For example, if variations in processing speed underlie individual differences, then changes in processing speed underlie developmental change. Equally, if individual differences are determined by differences in metacognitive functions, then changes in metacognitive functions underlie development.[1] Polar position 1 is most obviously revealed in psychometric conceptions of intelligence. Since, on this view, intelligence is best reflected by a score on an intelligence test, whether a particular score is interpreted in terms of individual differences (IQ: how the child's performance compares with the average score for the age group) or in terms of development (mental age (MA): the age for which that score would be average) depends on the subject group the researchers are investigating. Researchers interested in individual differences have focused on IQ, developmentalists on mental age. In either case the underlying metric is assumed to be the same.

In fact, it is interesting that attempts to squeeze the facts of individual differences and developmental change into this implicit framework have spawned a major debate: whether mentally retarded children develop differently or just more slowly (see Zigler and Balla 1977). This debate was provoked by the finding that MA-matched mentally retarded individuals are not cognitively equivalent to their normal controls, a finding not consistent with the common developmentalist notion that mental age reflects intellectual functioning during the developmental period, whereas IQ is simply an artifact of rate of development. Yet differences in IQ still predict cognitive differences, even when mental ages are equivalent. We will return to this issue in chapter 8. For the moment it is enough to observe that polar position 1, expressed in terms of the minimal cognitive architecture, would argue that if the speed of the basic processing mechanism is the primary causal variable in individual differences, then increasing speed will be the cause of developmental change (option 2 in figure 6.2).

Polar Position 2. Individual differences and developmental change are

---

[1] This is not to say that researchers who accept those particular models of individual differences or models of development explicitly recognize the connection between these two aspects of intelligence (for a start, they tend to be different groups of people); it is simply implicit in their theorizing.

caused, in principle, by unrelated processes. For example, Piagetian theory is explicitly unconcerned with individual differences, proposing that the processes of developmental change are *universal* and independent of those that might contribute to individual variation. While it may be the case that children of the same age and at the same level of development differ in their ability to acquire knowledge, changes in representational structure that occur with development constitute an altogether different process to that which influences individual differences in knowledge-acquisition within a stage. The development of intelligence entails a change in the structure of thought and representation. It is not simply a matter of acquiring new knowledge, a process undoubtedly subject to individual differences. Within the minimal cognitive architecture, such a proposition (the independence of developmental change from individual differences in knowledge-acquisition) can be afforded only by the hypothesis that cognitive development is the result of changes in the availability of modules (option 3 in figure 6.2), because only knowledge furnished by modules is unconstrained by the speed of the basic processing mechanism.

Clearly, then, the issue of whether intelligence and development are varieties of the same 'thing' hinges on two central issues:

1   the extent to which development reflects abrupt and universal changes in the structure of knowledge, as opposed to the more continuous variation seen between individuals;
2   the extent to which individual differences in intelligence influence cognitive development.

I will deal first with the issue of qualitative change in development.

Continuity or Qualitative Change in Development?

Piaget and Cognitive Development

Why should we consider Piagetian theory at all? Although Piagetian theory has been severely criticized since the 1970s and many of its empirical propositions regarding the intellectual competence of young children have been called into question, most contemporary theories of cognitive development orient themselves with respect to many Piagetian principles (for example, Carey 1985, Case 1985, Halford 1982, Karmiloff-Smith 1991, Kohlberg 1969). Piaget set the agenda regarding what theories of cognitive development should seek to explain. Thus, conceptual change in domains such as the understanding of space, causality, and quantity are still the focus of current research. It is safe to say that, as yet, nothing has replaced Piagetian theory as a general theory of cognitive development. It is incumbent, then, on any theory of cognitive development to address the important issues that Piaget raised. Piagetian theory also serves as the classic example of a theory that postulates qualitative changes in the structure of knowledge with development.

There are four central features of Piaget's theory of cognitive development.

(1) The most obvious feature is its emphasis on cognitive development as a progression through qualitatively different *stages* of thought. Although these stages are hierarchical, with each new stage building on its predecessor, the way in which knowledge is represented varies radically from one stage to the next. Each stage represents a new mode of thought, with a different logical structure, one which applies to many different cognitive domains, supplanting older versions. This gives Piagetian developmental theory a domain generality that is not an inherent component of a simple knowledge-acquisition theory, a quality that we will find useful.

(2) Piaget concedes that there are structural constraints on cognition, but argues that these constraints derive solely from the structure of our sensory transducer systems. There are no constraints imposed by cognitive systems themselves. So our most basic concepts are constrained – for example, our intuitions about space are influenced by the nature of our perceptual system – but intelligence, or the cognitive structures created by the *invariant functions of thought*, is not so constrained. Thus:

> Our Perceptions are but what they are, amidst all those which could possibly be conceived. Euclidean space which is linked to our organs is only one of the kinds of space which are adapted to physical experience. In contrast the deductive and organising activity of the mind is unlimited and leads, in the realm of space, precisely to generalisations which surpass intuition. (Piaget 1953, p. 4)

Piaget's view regarding the nature of cognition bears, in this context, a remarkable similarity to a particular aspect of Fodor's (1983) notion of modularity discussed in chapter 4. According to this, the structure of our sensory organs (*input modules*) constrains our basic concepts (offers particular kinds of representations of the world), but intelligent thought (*central processes*) is unlimited (indeterminate). But if thought is unconstrained, why should there be a universal sequence of developmental change? Clearly, for Piaget, development is not a case of 'Anything goes'.

For Piaget, although thought *within* a stage is unconstrained – that is, we can think anything we like (akin to Fodor's view of belief fixation in central processes) as long as the structure of the thought conforms to the logical form of the appropriate developmental stage – this does not mean that there is, consequently, a lack of constraint on developmental change. The constraint on development is a function of both the range of environmental circumstances to which individuals are exposed and the *logical* demands of the structures of thought designed to assimilate and accommodate them. This suggests that the aspect of intelligence that is unconstrained is that which refers to variations within the same logical form (within-stage variation). It also suggests that this variation is probably independent of the *changes* in logical form that are the hallmark of cognitive development. Given this interpretation, the third feature of Piaget's theory which is of interest to us comes as no surprise.

(3) Piaget is unconcerned with individual differences. The major focus for him is universal change in the structure of knowledge through development. Piagetian theory thus implies that all children pass through these different stages of thought, although little is said about the precise timing of these changes for individuals. Nor are individual differences in the degree of elaboration of knowledge within stages given much import. What is crucial for Piaget is that the sequence of stages should be invariant across individuals. So it cannot be the case that some children attain formal operational thought before they have attained concrete operational thought. However, the homogeneity of stages has been called into question (and therefore the whole concept of 'stage') by the fact that children can show concrete operations within one knowledge domain – for example, number – while remaining pre-operational in another – say, in conservation of length. We will return to this below.

(4) Despite the radical changes that cognitive structures undergo during development, Piaget leaves the door open for a distinction that we have found crucial so far in this book. Although 'intelligence' for Piaget is, of course, a property of these changing knowledge structures, there is also a sense in which intelligence can be regarded as the property of the *process* of knowledge acquisition. It is interesting that in this latter sense Piaget may consider intelligence to be unchanging.

The solution . . . is precisely to be found in the distinction between variable structures and invariant functions. Just as the main functions of the living being are identical in all organisms but correspond to organs which are very different in different groups, so also between the child and the adult a continuous creation of varied structures may be observed although the *main functions of thought remain invariant.* (Piaget 1953, p.4; emphasis added).

The invariant functions are the processes of *assimilation*, whereby new information is brought within the scope of existing cognitive structures, and *accommodation*, whereby structures change in order to eliminate mismatches between predicted and actual events. The principle that guides the cycles of assimilation and accommodation is *equilibration*, a homeostatic principle which requires that the state of the child's cognitive system be adapted to its world. So for Piaget, intelligence is a function of the degree of adaptation of cognitive structures to the world, but the processes of thought which underlie intelligence are themselves unchanging in development. I now want to flag a possibility for later discussion. The developmental processes of assimilation and accommodation look, to my eyes, like descriptions of knowledge-acquisition procedures that take place *within* stages characterized by particular logical properties of thought. Changes between stages may be due to quite different kinds of developmental processes.

Implications of the Piagetian view for the developmental dimension

There are a number of advantages to maintaining the core of Piagetian theory within the developmental dimension.

First, many developmental theorists accept that the core of development, as opposed to, say, learning, is a qualitative change in cognition. Children of different ages have different concepts and think in qualitatively different ways.

Second, a Piagetian-like theory, with an invariant sequence of change in the structure of cognition with its inherent domain generality, can easily accommodate our major developmental regularities. Sweeping changes in logical power brought about by the move from one stage of thought to another could explain why we see a fairly regular developmental pattern across different cognitive domains.

Third, the proposition that the processes underlying individual differences are independent of the processes underlying development would make the stability of individual differences during development unremarkable.

How could the developmental dimension of the minimal cognitive architecture accommodate the Piagetian thesis of qualitative change in development and the independence of this change from individual differences? There is but one way: the maturation of modules could account for increases in competence which may appear to create a new 'stage' in development, and, since modules are unconstrained by the speed of the basic processing mechanism, the increase in competence should be universal and should occur irrespective of individual differences in intelligence. I will from now on refer to this as the 'modular change' option for the developmental dimension of the minimal cognitive architecture (option 3 in figure 6.2).

But how plausible is it to map Piagetian qualitative change onto the maturation of modules? At first blush, there is a problem with the modular change option that appears to destroy one of the principal advantages of the attempt to accommodate Piagetian theory. I argued in chapter 4 that modules are function-specific knowledge-acquisition mechanisms; that is, that they generate only one kind of knowledge. This may restrict their usefulness as a general influence on the development of thought – the very attribute that changes between Piagetian stages could be invoked to capture. However, the specificity of modules, in the sense that they afford a particular kind of representation, may not be, in fact, relevant to their generative capacity. Let me explain.

If the new kind of representation afforded by a module – say, a linguistic one – is generally useful for many kinds of knowledge-acquisition, then we would expect its appearance to have profound stage-like qualities on the nature of thought. The ability to represent the world linguistically, a competence afforded by a specific module, would allow us to entertain *linguistic* propositions which, in turn, would have a generalized effect on thought. In a sense, new modules may afford new expressions for a language of thought. So modules may have more generalized effects on cognition than may be obvious at first. But maybe the weaker generality of *modular change* as compared with Piagetian stages is, anyway, an advantage when seen in another light.

There is considerable slop in the data regarding stage changes in cognition, slop that the Piagetian notion of *décalage* was designed to overcome. For classical Piaget, development should be more discontinuous and domain-general than it often appears to be. We rarely see step-like changes in development. And sometimes apparent stage changes in one domain do not spill over into others. For example, some children demonstrate conservation of number, but not volume. Piagetians have explained these discrepancies in two ways:

1   Horizontal *décalage* (the lag between changes in stage in different knowledge domains) across subjects would make performance when averaged across domains seem more continuous than it is. But what is *décalage*?

2   The step-like change in development that should appear in response to a major restructuring of knowledge would, in practice, be blurred by the process of assimilating new information to changed structures, a process which would presumably take some time and would also produce smaller, more gradual improvements in cognitive performance. For example, if we regard logical operators as functions that require certain kinds of arguments, the form of the data structures that have been laid down for any particular knowledge domain will determine the ease of application of a new logical operator.[2] The modular change option almost presupposes such a process. A new competence for linguistic representation, for example, will take some time to influence a variety of knowledge domains, and will influence some more easily than others.

In this sense, the notion of modular change followed by knowledge elaboration fleshes out the Piagetian formulations of both *décalage* and assimilation. Together with the greater emphasis on domain specificity inherent in the notion of module, as compared with a Piagetian 'stage', the process of knowledge elaboration explains why development appears less discontinuous than a qualitative-change theory suggests it should.

Two Kinds of Developmental Change?

The option, for the developmental dimension, of modular change can capture the qualitative change so central to Piagetian theory, but, being more domain-specific (depending on the module), it could leave some aspects of cognition unaffected. This does away with the need for the special pleading inherent in the concept of *décalage*. In addition, the new mode of representation offered by a newly acquired module – for example, a language-acquisition device will still have to be utilized in more general and idiosyncratic knowledge bases. *This* process will obviously be constrained by the speed of the basic processing mechanism and the power of the specific processors. In Piagetian terms, assimilation will be the process most obviously related to individual differences.

---

[2] This notion is part of Karmiloff-Smith's representational redescription hypothesis. I would like to thank John Morton for bringing it to my attention.

We are left, then, with the proposition that there are two kinds of developmental changes.

(1)   Major developmental changes occur with the maturation of a new module. Since modules are unconstrained by the speed of the basic processing mechanism, such developmental changes will be unrelated to individual differences in intelligence. At least in the case of language acquisition, such an idea is not without precedence. Developmental psycholinguists have long debated the relationship between cognitive development and language acquisition. The *cognition hypothesis* argues that there are cognitive or conceptual prerequisites for language development; although in its sophisticated forms, it has always recognized that some aspects of language development are independent of general cognitive processes (Cromer 1974).

The evidence on whether language development depends on general cognitive processes or whether, by contrast, it is language that drives cognitive development is a minefield of interpretative difficulties. Certainly there are individual differences in language development and certainly most mentally retarded children are also linguistically backward; but correlations do not necessarily indicate causal relationships. The existence of cases like D.H. (see chapter 4), whose language has developed to an advanced state despite severely retarded cognitive development, suggests that language development and cognitive development can be independent of each other. Cromer, who has provided the definitive analysis of the cognition hypothesis, states that 'correlations based on the development of normal, unimpaired individuals are never decisive. What cases such as D.H. seem to show is that general cognitive mechanisms are neither necessary nor sufficient for the growth of language' (Cromer 1991, p. 135).

Also consistent with the modular development thesis are studies which claim that many linguistic milestones are better predicted by the rate of attainment of *motor* milestones (which are likely to be indicative of biological maturation) than by IQ differences (Lenneberg *et al.* 1964).

Modular change, then, would involve major, and universal, discontinuities in underlying *competencies* or, perhaps, in the language of thought (Fodor 1975). Of course, modules will not mature at the same rate for everyone, but their rate of maturation will be unrelated to individual differences in intelligence. The exception to this would be in pathological cases where central nervous system disorders cause both a delay in the maturation of modules and, coincidentally, subnormal levels of intelligence.

(2)   The second developmental process is one of *elaborating* knowledge bases using this new representational competence. This second developmental change would be related to individual differences in intelligence, since idiosyncratic knowledge acquisition utilizing a new module would be subject to the constraints of the basic processing mechanism and the specific processors. So, for example, suddenly having access to linguistic representations will result in a discontinuity in development that is unrelated to individual differences in intelligence. However, elaborating one's knowledge of the world by using this new competence – for example, by building vocabularies – will be constrained by the basic processing mechanism and the specific processors. Thus individual differences in intelligence will have an influence on the elaboration of knowledge, but not on the acquisition of new competencies.

The theoretical work of Piaget on discontinuity in development is roughly consistent with a modular change option for the developmental dimension. The reader should not be too troubled if Piagetian theory seems somewhat distorted by the attempt to squeeze it into the mould of the minimal cognitive architecture. Piaget's theorizing was, as we have noted, unconcerned with individual differences and hence unconstrained by this dimension of cognition. It would be surprising, therefore, if it slotted into the new theory with no trouble at all! Whether it is worth all the squeezing depends, of course, on whether individual differences in intelligence do have a systematic influence on cognitive development. But before we go on to consider the evidence regarding the influence of individual differences in intelligence on development, let us first evaluate some more evidence regarding the issue of continuity and qualitative change.

Infant Intelligence

The Piagetian theory of radical changes in the structure of knowledge has been used to explain the failure, despite many attempts, to find measures of infant intelligence that will predict later psychometric performance (Bayley 1933, McCall 1976, Lewis *et al.* 1986). The argument is, simply, that because development involves discontinuity (qualitative change) in cognition, there is no reason to suppose individual differences during one stage would predict individual differences during any other. However, it may be that this evidence for discontinuity is an inevitable consequence of the Piagetian framework. We can see this more clearly by examining what constitutes a standardized test of infant intelligence.

The major infant intelligence tests – for example, the Bayley test – assess infant behaviour against developmental milestones. The behaviours assessed are typically related to sensory-motor performance. Does the baby track an object with its eyes when it is moved in its visual field? Will it reach for objects? Will it reach for an object that has been hidden underneath another one? And so on. It is a well established fact that there is very little relationship between an infant's developmental status, as indicated by performance on such tests, and its later IQ. More intelligent children are not faster developers in infancy (Brooks and Weinraub 1976). Much of the evidence cited for discontinuity in development is based on this fact. However, with the hindsight that our new theoretical perspective gives us, we can see that there is an underlying assumption that makes it unlikely that this research could have found any continuity between infant measures of intelligence and later IQ.

The early research on infant 'intelligence' assumed that whatever 'intelligence' is, it changes with development. The obvious strategy, then, if one is looking for correlates between infancy measures and later differences in intelligence, is to go for those measures that show the greatest developmental change. This neglects the possibility that *changes* in cognitive abilities during development may be unrelated to individual differences. However, infant intelligence research so far seems to indicate that, for this period of development at least, these two dimensions of cognition are indeed unrelated. It seems that much of development in infancy is concerned with perceptual and motor systems, and that these systems are unrelated to the cognitive processes underlying intelligence. Thus the behaviours that reflect individual differences in intelligence may not be those that undergo the most obvious developmental change. It was only when cognitive processes were investigated *for their own sake* that the discovery of a relationship between individual differences in infant cognition and individual differences in later childhood even became possible (Fagan and Singer 1983, Bornstein and Sigman 1986).

Fagan and McGrath (1981) reanalysed data gathered by Fagan (1971, 1973, 1976, 1977) to assess the effects of delay on immediate recognition memory in infants. Fagan used a standard technique, *paired comparison*, for assessing recognition memory (Fantz and Nevis 1967). Infants were familiarized with a stimulus, usually a slide of an object or a shape, for somewhere between 30 seconds and 2 minutes and were then presented in a test phase with the familiar stimulus paired with a novel one. The amount of time the infant spent looking at each stimulus was recorded during the test phase. The extent to which the infant looked at the novel, as compared with the familiar, stimulus was taken as an index of *novelty preference*. Fagan and McGrath (1981) reported significant correlations between measures of visual novelty preference at 4–7 months and IQ scores at 4 years old in one group of children (r = 0.37, n = 54) and at 7 years old in another group (r = 0.57, n = 39). Thus, infants who showed a greater preference for the novel stimulus turned out to have higher IQs (Fagan 1984; O'Connor *et al.* 1984), and novelty preference was significantly lower in babies with a high risk of being mentally retarded (Miranda and Fantz 1974).

The same result was found by Lewis and Brooks-Gunn (1981) using a completely different testing paradigm, *habituation*. Infants were presented with the same stimulus over a number of trials, and the duration of their visual fixation on the stimulus was noted. Infants were found to habituate to repeated presentation of the same stimulus; that is, their fixation times decreased until they could be said (on some objective criterion) to have habituated. At this point the infants who constituted the experimental group were presented with a novel stimulus, which differed from the familiar stimulus along some dimension of interest. The control group was given another trial with the familiar stimulus. The assumption of the paradigm was that if the difference between the novel and the familiar stimulus was detectable, then the experimental group would show increased looking by comparison with the control group. To the extent that they did, they could be said to have *dishabituated*. In this context, the degree of dishabituation can be taken as an alternative index of novelty preference. Lewis and Brooks-Gunn (1981) tested 79 infants at 3 months, using a modified habituation procedure, and tested the same infants on the Bayley test of infant development at 3 months and again at 24 months. Consistent with previous findings in the literature, there was no significant correlation between the Bayley test scores for the two testing periods. However, the correlation between dishabituation measures at 3 months and Bayley test scores at 24 months was 0.52 and 0.4 respectively for two subgroups in the sample, indicating that this measure of infant novelty preference could predict standardized test performance at least by the end of the first two years of life.

New measures of cognitive, rather than sensory-motor, processes have discovered some continuity between infant and later intelligence. How has this been explained?

Bornstein and Sigman (1986) discuss two principal models of continuity, the first of which, the continuity of underlying processes model, can be interpreted in three ways:

1   Continuity is due to a shared reliance on g. This is regarded as an unlikely candidate by Bornstein and Sigman, principally because the best correlations are found with verbal IQs, and usually not with other measures of cognitive abilities.[3]

2   Continuity is due to a shared reliance on what Bornstein and Sigman call 'mental representation'. By this they mean the collection of specific processes involved in any information-processing task (stimulus encoding, short-term memory, response selection, and so forth). Mental representation is seen as a more likely basis for continuity than general intelligence but this forms the extent of its evaluation by Bornstein and Sigman.

3   Continuity is due to a stable, but non-cognitive influence such as arousal, motivation, or attitudinal variables. Although Bornstein and Sigman regard these non-cognitive traits as plausible contributors to continuity, they acknowledge that measures of these traits in childhood show a fair degree of independence of the measures of intelligence that are predicted by the infancy tests.

The second of Bornstein and Sigman's models is that of continuity of developmental status. This argues that cognitive abilities may be independent, but that they may also develop in parallel within each child, and that the rate of development they follow may differ between children as a function of intelligence. Thus the infant tests are useful predictors because they reflect a relatively stable 'developmental standing' (Bornstein and Sigman 1986, p. 263). Bornstein and Sigman point out, however, that this fails to account for the lack of predictability shown by traditional infancy tests, which could be regarded as measures of developmental status. They do not mention that, in any case, an explanation based on continuity of developmental status would simply shift the burden of explanation to the question: Why do children have different rates of development?

There are, then, a broad range of possible explanations, albeit of varying plausibility, of why measures of looking time in infancy could predict later differences in measured intelligence. It is useful, given such a broad range of explanations, that the theory of the minimal cognitive architecture should make clear predictions about the likely basis of the continuity between cognitive measures in infancy and measures of intelligence in later childhood. If infancy measures can predict a sub-

---

[3] This ignores the fact that there are differential reliabilities in children's intelligence test performance, with verbal tests being the most reliable. The more reliable the test, the more likely it is that significant relationships with other variables will be found. In addition, verbal tests are known to be more g-loaded than performance tests.

stantial proportion of the variation in childhood IQ scores, then, since, according to the theory of the minimal cognitive architecture, the speed of the basic processing mechanism is the major determinant of individual differences in intelligence, it is likely that the infancy measures also reflect the speed of the basic processing mechanism. So, is it possible that the infancy measures are indexing differences in processing speed? This explanation looked implausible until very recently.

The early studies of Fagan led to the obvious conclusion that the key to the continuity lay in understanding novelty preference. R. J. Sternberg (1981) interpreted the infancy data as supporting his notion that non-entrenched thinking is a key component of intelligence. Thus the infant who embraces novel events will be the young child who does likewise and who will acquire, as a result, a more varied knowledge base. This more varied knowledge base will be reflected in higher IQ scores. Fagan and Singer (1983), although of the opinion that the search for continuity and the search for *g* amount to the same thing, also interpreted the data as supporting the idea that novelty preference indexed the full range of an infant's information-processing skills. These earlier theories thought that the key to understanding the continuity was to be found in the way that novel stimuli are processed. However, there is a final twist to this tale which casts doubt on the idea that it is something peculiar to the processing of novel stimuli that generates the relationship with later IQ.

By contrast with the burgeoning research using the paired comparison paradigm, the research focusing on the dishabituation/later IQ relationship was beset with problems, both methodological (McCall 1979, 1981) and in their failure to replicate. More recently, however, the strongest relationships yet reported between infancy measures and later IQ have been found to be time-based measures, usually taken during habituation studies (Rose *et al.* 1986, Cohen 1981). In addition, the indices that show the strongest relationships with later IQ differences, such as duration of first fixation ($r = -0.85$) and average fixation duration ($r = -0.76$), are among the most stable and reliable measures of infant looking behaviour during the first year of life (Colombo *et al.* 1987).[4] It is the infants who look for the *shortest* time, particularly on the first presentation of the stimulus, that turn out to have higher measured intelligence in childhood. Interest in novelty, *per se*, seems to be a poor predictor, because *dishabituation*, by contrast, shows little predictive validity. Indeed, an explanation based on the supposition that the

[4] Both correlations reported are between habituation measures taken in infants under 6 months of age and British Ability Scales scores when retested at 4.5 years (Rose *et al.* 1986).

more intelligent infant has a greater interest in novelty might predict that it would be the infants who looked for the longest time when presented with the first habituation trial (given that this is most certainly a novel event) who would turn out to be the most intelligent.

If we assume that shorter looking times index faster information-processing, we can now make the connection between the mechanisms of the minimal cognitive architecture and the aspects of infant looking times that predict later IQ. The hypothesis is that the infancy measures that predict later IQ do so because both looking times in a habituation study and intelligence test performance indirectly measure the speed of the basic processing mechanism. But how can this account for the original novelty preference effect reported by Fagan and McGrath (1981)?

If we hypothesize that for the fixed familiarization period in the Fagan and McGrath study the faster processors will have extracted more information from the stimulus than the slower processors, then the novel stimulus will be more novel for those infants who process information faster. If interest in a novel stimulus is a function of the *degree* of familiarity with the original stimulus, then faster processors will show increased novelty preference. Some evidence for this hypothesis comes from studies which show that novelty preference is manipulated both by familiarization time and by stimulus complexity. Some studies have found that younger children prefer a familiar to a novel stimulus (Wetherford and Cohen 1973, Greenberg *et al.* 1970). Others have shown that the preference for a familiar or a novel stimulus is a function of stimulus complexity, which is itself age-related (Hunter *et al.* 1983). Similarly, preference for novelty tends to result from a long familiarization period, whereas preference for the familiar seems to arise from a short familiarization period (Cohen and Gelber 1975, J. M. Hunt 1974). It seems plausible, then, that the original relationship between novelty preference and later IQ was mediated by the speed at which stimuli were processed in the familiarization period. With a different set of stimuli or a shorter familiarization period, a familiarity rather than a novelty preference might have correlated with later IQ. The real predictive variable is not novelty preference, but processing speed.

Implications of infant intelligence for the developmental dimension

The fact that there are measures in infancy that predict later differences in measured intelligence, coupled with the possibility that they may do so because of a shared basis in the speed of processing, has a number of implications for the developmental dimension of the cognitive architecture.

The sense in which Piaget invoked discontinuity in development can be captured, as we have seen already, by the proposition that the major developmental changes are provoked by the maturation of new modules. Clearly, infant intelligence tests which measure individual differences based on behaviour typical of what Piaget termed the sensori-motor period are unrelated to the individual differences as measured by conventional intelligence tests some years later. This is what would be expected if development in the sensorimotor period is based on the maturation of modules. Modules (route 2 knowledge-acquisition) are the basis of the discontinuity in development. Continuity in development, on the other hand, is a continuity or stability of individual differences. And individual differences are based on differences in the mechanisms underlying route 1 knowledge-acquisition, particularly the basic processing mechanism. The continuity seen between infancy and childhood measures of intelligence is based on the continuity in the speed of the basic processing mechanism. In sum, qualitative changes in thought and representation in development are due to modular changes in route 2 knowledge-acquisition, and continuity is due to the unchanging mechanisms underlying route 1 knowledge-acquisition (which are the basis of individual differences in intelligence). There is, therefore, no conflict in there being marked continuity and marked discontinuity in development. Further, this means that development and individual differences, being related to different kinds of knowledge-acquisition mechanisms, are fundamentally orthogonal.

Two Kinds of Development Revisited

I should be careful not to overemphasize the independence of these two dimensions of intelligence. Although the developmental dimension is primarily dependent on route 2 knowledge-acquisition (the maturation of modules), there must be a secondary aspect of intelligence that is influenced by individual differences in intelligence. Individual differences are based on route 1 knowledge-acquisition mechanisms. In particular, the speed of the basic processing mechanism constrains the complexity of thought and consequently the complexity of knowledge that can be acquired via this route. I called this secondary kind of development 'knowledge elaboration'. The last major body of evidence relevant to the specification of the developmental dimension relates to the relationship between individual differences and developmental change. Given the story so far, we should focus on the extent to which any influence can be characterized as being confined to knowledge elaboration, as opposed to the acquisition of new competencies afforded by the maturation of new modules.

The Influence of Individual Differences on Development

The Developmental versus Difference Controversy

No one disputes that the retarded are cognitively handicapped; that is, that when compared with non-retarded children of the same age, they show poorer performance on every task that can reasonably be claimed to reflect reasoning and thinking skills. If individual differences in intelligence influence development, then we might expect that children with low IQs would develop differently from children with normal IQs. However, simply establishing that low IQ leads to delayed development is trivial and tautological: children with lower IQs must have lower mental ages than their higher-IQ same-age peers, as this is a necessary consequence of the way that IQs and mental ages are derived. But does IQ influence only rate of development?

Developmental level is usually taken to be synonymous with mental age. Given this, we can ask a more specific question: Does having a lower IQ (that is, being mentally retarded) still have consequences when children are matched for developmental level – that is, mental age? The answer seems to be an unequivocal 'Yes', but the explanation for why this should be so has generated a substantial debate: the *developmental/difference* controversy (Zigler and Balla 1977).

The developmental view maintains that the retarded develop the same kinds of cognitive abilities as normal children, albeit more slowly and perhaps with a lower ceiling. But on this view, when retarded and normal children of the same developmental level (that is, mental age – MA) are compared, there should be no difference between them in cognitive performance. The fact that there is a difference is attributed to non-cognitive factors, such as institutionalization, motivation, and learned helplessness, associated with being retarded.

The difference view, by contrast, maintains that the retarded have a cognitive deficit compared with normal children even when apparently matched for developmental level. Matching for MA is not thought to compensate for this basic cognitive deficit, because the retarded achieve their equal MA via a different developmental route. In other words, for difference theorists, IQ has cognitive consequences over and above those captured by MA.

The evidence for the developmental view is uncompelling while that for the difference view is convincing. Spitz (1983) has produced a cogent argument suggesting that the developmental position is flawed both logically and empirically. For example, he points out that while it may be the case that motivational factors influence the cognitive performance of mentally retarded people more than that of the non-retarded, this has already been controlled for in the estimate of their MA, which should then, on the developmental view, *underestimate*, not overestimate, their underlying cognitive ability. Likewise, the main body of empirical evidence favouring the developmental hypothesis has been the finding that institutionalized retarded people do significantly worse on cognitive tasks than non-institutionalized retarded people. However, it is also the case that the former group tend to have lower IQs and that once IQ is controlled for, the effect of institutionalization disappears.

Consistent with the difference hypothesis, Spitz (1982) showed that retarded and normal children achieve equal MAs in very different ways. For example, the retarded are relatively poor on tests such as *Similarities, Definitions, Information, Vocabulary,* and *Word Fluency* from the Wechsler scales, but are relatively good on *Object Assembly* and *Picture Completion,* as well as items on the Stanford Binet scale such as *counting backwards* from 20 to 0 in 40 seconds, giving the current *date,* and calculating the *change* that should be given for certain purchases. Spitz concluded that on tests which require reasoning and abstraction, retarded subjects are still handicapped compared with their MA-matched controls; but that on tests which tap 'experience, rote performance and maturation' (Spitz 1982, p. 172), they do better than younger normal children. In other words, on tasks requiring complex manipulation of abstract information MA-matched retarded subjects still show poorer performance than normal children. Spitz terms this phenomenon 'MA lag', since retarded children perform more poorly on some tasks than they should, given their mental age. The retarded achieve their equal MAs with non-retarded groups by being relatively superior on other tasks, specifically those for which their greater age can compensate in terms of opportunities afforded for practice. Spitz goes on to show that the items on which the retarded are most handicapped are those which are psychometrically *g*-loaded. In other words, the *g* loading of a test in a test battery predicts the MA lag associated with that test. This is consistent with the idea that it is those tasks that tax the speed of the basic processing mechanism that are most affected by mental retardation.

Implications for the Developmental Dimension

Taking the perspective of the minimal cognitive architecture, low IQ must mean a slow basic processing mechanism. If matching for MA were equivalent to matching for speed of the basic processing mechanism, then any IQ differences between the groups should have no effect. But since IQ does seem to make a difference to the relative cognitive strengths and weaknesses of MA-matched low- and normal-IQ groups, and since the tasks that most reflect these relative strengths are those that are highly loaded on general intelligence, then it is reasonable to suppose that the speed of the basic processing mechanism is slower in retarded groups than would be predicted from their mental ages. It is, then, a short step to conjecture that the speed of the basic processing mechanism is unrelated to mental age in normal development; that is, it might not change at all and only ever reflects differences in IQ. If this is so, it *must* be the case that MA matching can never produce cognitive equivalence, but merely allows for knowledge acquired through greater experience to compensate, in some ways, for underlying differences in processing speed.

In addition, the variation in cognitive abilities contributed by variations in the development of knowledge is itself related to speed of processing. Individuals with higher processing speeds will have more variation in knowledge elaboration than individuals with lower processing speeds, because variation in experience will matter more. In effect, the person with fast processing is capable of extracting knowledge from the full range of environmental circumstances in which he or she might find him or herself. On the other hand, the person with slow processing can only make sense of the simplest experiences. Variation in knowledge elaboration will, then, contribute more variation to the abilities of the higher-IQ child. This might explain why the stability of IQ is greater for mentally retarded populations than for normal populations (Spitz 1982).

We have now arrived at a specification of the developmental dimension. Major developmental changes are due to the maturation of modules (route 2 knowledge acquisition). Because modules are unconstrained by the speed of the basic processing mechanism, this kind of developmental change will be independent of individual differences in intelligence. Such change will be related to biological age or its closest correlate, chronological age (CA). This 'pure' form of development may be easier to spot in the case of a developmental deficit, since it is likely to pass unnoticed in the normal run of events (except perhaps in infancy). So, for example, it is interesting that some autistic children fail 'theory of mind' tasks that children with mental ages several years younger pass with ease (Baron-Cohen *et al.* 1985) and that this capability may be underlain by mechanisms that are plausibly modular (Leslie and Frith 1990, Leslie 1991). The ability to understand whether others have mental states, while having profound consequences on cognition if absent, may not be an 'obvious' milestone in normal development.

A secondary developmental process is knowledge elaboration. Knowledge elaboration will depend primarily on the mechanisms underlying route 1 knowledge-acquisition – namely, the basic processing mechanism and the specific processors. This is the locus of the effect of individual differences in intelligence on development. Such an effect will manifest itself in a correlation between IQ and knowledge elaboration.

Mental age (MA) will reflect both these developmental processes, and remains the best psychometric estimate of 'developmental level'. Although IQ will predict knowledge elaboration, this secondary developmental process must be recursive. That is, what determines the current state of knowledge is not only the speed of the basic processing mechanism (approximated by IQ) and the facility it offers for elaborating new knowledge, but the previous state of knowledge (approximated by MA). In addition, modular changes (an approximate function of CA) may entail a reorganization of knowledge and may provide, perhaps, new expressions for a language of thought, the consequence of which will be an increase in the complexity of thought itself. Such an increase in the complexity of thought will, then, be most obviously correlated with mental age, and may constitute the core of what Piaget meant by cognitive development.

The specification of the developmental dimension is consistent both with data on individual differences *and* with some of the data and ideas on developmental change. Undoubtedly there will be developmental data that may appear inconsistent with the view I have sketched (although I am unaware of any glaring anomalies), but since the formulation is new, it is unlikely that the existing data base will provide a fair test. Such a test will have to await future research. Roughly put, cognitive performance and developmental change that is dependent on route 2 knowledge-acquisition (modules) will be related to chronological age; whereas cognitive performance and developmental change (knowledge elaboration) that is dependent on route 1 knowledge-acquisition will be related to IQ; finally, cognitive performance and developmental change that is dependent on the current state of knowledge will be related to mental age. As it happens, there already exist some data that touch on these predictions.

Piagetian Development and the Growth of Mental Age

In a series of studies Weisz and colleagues (Weisz and Zigler 1979; Weisz and Yeates 1981; Weiss *et al.* 1986) explored the developmental/difference controversy using both Piagetian and information-processing measures. They divided the developmental/difference controversy into two distinct hypotheses:

Hypothesis 1: *similar sequence*. The mentally retarded pass through the same developmental stages in the same order. Weisz and Zigler (1979) showed in a review of 31 studies that, with the exception of organically brain-damaged groups, the preponderance of evidence suggested that this hypothesis is correct.

Hypothesis 2: *similar structure*. When the mentally retarded are equated with normal controls for mental age (developmental level), they will have the same cognitive structure, structure being defined as the organization of thinking and learning processes. As we have seen, this hypothesis constitutes the kernel of the dispute. The developmental position argues for similar structure in MA-matched retarded and normal groups.

Weisz and colleagues investigated cognitive structure from two different theoretical perspectives: first, the Piagetian (Is the structure of knowledge the same in MA-matched retarded and normal groups?) and second, information-processing (Are the information-processing capabilities of the two groups similar?)

Weisz and Yeates (1981) had previously reviewed 30 studies comparing retarded and normal children of equal MA on Piagetian tasks. The review covered 104 tasks in all, including tests of conservation of number, quantity, mass, and length and tests of seriation, number, transitive inference, classification, moral judgement, and perspective taking. They found that on only 4 per cent of the tasks did the mentally retarded show any superiority to their younger control group, whereas on 24 per cent they showed a significant MA lag. However, on 72 per cent of the tasks there was no reliable difference, which may be taken as evidence in support of the similar structure hypothesis, Piagetian style. As Weisz and Yeates (1981) pointed out, however, the similar structure hypothesis is also the null hypothesis. They reasoned, therefore, that the number of comparisons showing significant differences between the groups should be equal to, or less than, the total number of comparisons multiplied by the alpha level of significance (0.05). Chi square tests showed that when this was done for the subset of studies in which the mentally retarded could be identified as organically impaired, the null hypothesis could be rejected. These groups do produce significantly more differences than would be expected by chance. On the other hand, when the mentally retarded groups with no obvious organic impairment are compared with their MA-matched normal controls, the null hypothesis cannot be rejected. Thus Weisz and Yeates (1981) concluded that, in terms of performance on Piagetian tasks, MA-matched retarded and normal children have a similar cognitive structure, except where the retarded group have an identifiable organic impairment.

Weiss *et al.* (1986) applied the same logic to a review of performance on some 59 'information-processing tasks', including measures of attention, memory, concept naming, short-term memory processes, and iconic memory. On 44 per cent of these tasks the retarded groups performed significantly more poorly than the controls. This exceeded the number of such differences that could be expected by chance by a factor of ten. Thus, whereas the distribution of comparisons on Piagetian tasks did not differ significantly from the normal distribution predicted by the null hypothesis, and therefore by the developmental position, this is not the case for the 'information-processing tasks'.

Implications for the Developmental Dimension

This analysis of a number of well-controlled studies gives some support to the idea that certain developmental changes are relatively immune to individual differences in intelligence as measured by IQ. It seems that the kinds of universal changes in conceptual structure that are the focus of Piagetian theory are, indeed, universal. Thus, matching for mental age does produce a kind of structural equivalence. Certainly Piagetian tasks are relatively uninfluenced by IQ in comparison to tasks which tap on-line information-processing capacities, for which the lower IQ of retarded groups seems to produce significantly poorer performance even when matched with normal children for mental age. Piagetian similar structure was found in MA-matched groups, and if, as is likely given other research that demonstrates correlations between MA and Piagetian tasks (Humphreys and Parsons 1979, Inman and Secrest 1981, Carroll *et al.* 1984), such a structure is more correlated with MA than CA, then this suggests that classical Piagetian development is not modular (or it would be more CA-dependent) and depends on knowledge prerequisites.

There can be little doubt that with regard to many conceptual changes in development the retarded are, at the very least, delayed, and again we see evidence for two developmental processes. Major changes (discontinuities) in cognitive structure are brought about by the acquisition of new modules. This happens independently of individual differences in intelligence, and is probably chronological age-dependent. The *elaboration* of these new competencies in new conceptual structures, however, depends on the general state of knowledge: how much and what we know influences conceptual change. How much and what we know is, in turn, dependent on two things: the efficiency of our knowledge-acquisition mechanisms and how long we have been around. For this reason, mental age, being a cocktail of both the efficiency of knowledge-acquisition mechanisms and the amount of experience with which those mechanisms have been fed, is the best summary measure of the state of cognition. What equating mental age with developmental level obscures, however, is the fact that these two influences on knowledge-acquisition suggest fundamentally different kinds of acquisition profiles for groups differing in intelligence. And, indeed, when we look at the information-processing capacities of the retarded, we find them to be deficient by comparison with those of their normal MA-matched peers.

This review of the developmental profiles of retarded versus non-retarded children has generated a supplementary hypothesis to the developmental dimension. I argued above that the continuity in development is a continuity or stability of individual differences, primarily the speed of the basic processing mechanism. The fact that equating for mental age between retarded and non-retarded groups does not equate for cognitive performance, particularly on measures of information-processing capacities, suggests that IQ, and not MA, is the better predictor of the speed of the basic processing mechanism. In turn, this leads to the hypothesis that the speed of the basic processing mechanism does not change with development. What changes is the content and organization of knowledge, which depend on both the maturation of modules and the processes by which knowledge is elaborated. Certainly, unchanging processing speed would give rise to the substantial stability in IQ, particularly in lower-IQ groups, that is a major item on our explanatory agenda for a theory of intelligence and development.

This new hypothesis clashes with a great deal of research which seems to indicate that, far from being a dimension orthogonal to cognitive development, processing speed, or capacity, is the very *motor* of cognitive change.

Does Changing Speed of the Basic Processing Mechanism Cause Developmental Change?

Neo-Piagetian theorists claim that increasing capacity, or speed, of processing is the causal variable in developmental change (Pascual-Leone 1970, Halford 1985). This claim is equivalent to option 2 in figure 6.2. Although these theorists are, in the main, agnostic about individual differences, their general position would be enthusiastically championed by an unlikely group of supporters. If individual differences can be attributed to variations in some *low-level* aspect of cognition, where high speed, or efficiency, equals high intelligence, then the parsimonious assumption is that the development of intelligence is caused by increasing this parameter. So in this case, the increase in speed, or efficiency, that occurs with age *causes* cognitive development. Thus, the mechanisms underlying individual differences and those underlying development are considered to be the same. Theorists who attribute IQ differences to variations in low-level processes (Jensen 1982, Eysenck 1988) and cognitive developmentalists make unlikely bedfellows.

Can we invoke the notion of increasing speed of processing to account for developmental change, just as we have used the notion of variations in the speed of processing to account for individual differences?

Processing Speed, Capacity, and Developmental Change

Although neo-Piagetians, whose views are best exemplified by the two major theories of Case and Halford, take the line that development is accompanied by increasingly complex knowledge structures and reasoning abilities (if they didn't, they would not be Piagetian), they attribute some of the responsibility for this to an independent process related to cognitive *capacity*.

For Case (1985) development is best viewed as a change in the ability of the child to assemble executive control structures for solving different classes of problem. These control structures have three separate components: (1) a representation of the problem situation, (2) a representation of the problem objectives, and (3) a representation of the problem strategy. Case emphasizes that control structures for solving dissimilar problems are very similar in their underlying form, and that new control structures are superstructures composed of more elemental structures. Clearly, when a new superstructure is assembled, it will constitute a new unit of thought, and, depending on the scope and sophistication of the superstructure, it might even be regarded as representing a qualitative change in thought, in the older Piagetian sense.

In Case's theory there are four major levels, or stages, of development in control structures: sensorimotor, relational, dimensional, and vectorial, the details of which need not concern us here. Within each of the four stages there are four substages of development: operational consolidation, unifocal co-ordination, bifocal co-ordination, and elaborated co-ordination. This all sounds very Piagetian, so far, but now we diverge.

Case supposes that the assembled control structures utilize some kind of working memory system (Baddeley 1986). This system has associated capacity restrictions, both in a central executive and in a short-term memory store. In order to solve problems, computations (determined by particular control structures) must be performed in the central processing space, and intermediate products of these computations must be kept in the memory store. The total capacity of the system is a function of the *operating space* available for computation and that available for *memory storage*.

Each of the four major stages utilizes a certain operating space in the central executive, and each of the four substages imposes an increasing load on short-term storage space. One of the variables constraining stage transitions is the capacity requirements of a particular control structure. So, for example, Case argues that a particularly important transition takes place between 4 and 6 years of age, a period which marks the change from relational to dimensional control structures. Case believes that below the age of 4, children cannot execute a dimensional operation, such as enumeration, in the central operating space and simultaneously hold the product of a previous operation in the short-term memory store. In other words, any task that requires the comparison of two dimensions for its successful solution (for example, comparing the number of two sets of objects on a balance beam *and* their distance from the fulcrum to determine which side will go down) is impossible for the pre-dimensional child, because the demands of the dimensional control structures exceed the available capacity of the system. But what is the capacity change between 4 and 6?

Pascual-Leone (1970), whose theory formed the starting point for Case, hypothesized that *M-space*, or central processing space, increases in development. Case, although in sympathy with the general framework of Pascual-Leone, believes that the total capacity of the system remains unchanged throughout development, but that the operating space required by a particular structure decreases, freeing more capacity for short-term storage. Why do the operating space demands of a structure decrease? Case believes this is because particular structures are processed more 'efficiently', probably because of increased myelinization of neural pathways.

In sum, Case, like Piaget, proposes that higher-order cognitive structures involve the hierarchical integration of less complex lower-order structures. However, whereas Piaget regards the transition from lower to higher stages in development as dependent on an internal process that is logical in nature, Case believes that there is an increase in the efficiency with which control structures are executed which frees capacity for short-term memory storage. This then allows, for example, the results of dimensional computations (like x is bigger than y) to be held in memory to be used in future computations. This greatly increases the cognitive performance of the child.

Halford's theory (1985, 1987) is in much the same spirit as Case's, arguing that changing capacity demands are the motor of developmental change. The major difference is that Halford, unlike Case, believes that the capacity available to the child increases during development, rather than that the capacity demands of early developed cognitive structures decrease.

Just as the notion of control structure is central to Case's theory of cognitive development, so the notion of structure mapping is central to Halford (Halford and Wilson 1980). Halford argues that any cognitive task can be described in terms of the mapping requirements between the task and an internal representation. The representation of knowledge can be given a structural description. A knowledge structure comprises a set of elements with a set of relations, functions, or transformations defined over these elements (Halford 1987). The structure mapping is a rule for assigning the elements, relations, functions, and transformations of one representation to another. Such mappings form the basis of many different kinds of reasoning, such as analogical and inductive inference.

There are three levels, of increasing complexity, of structure mapping between representations. To understand these different levels, imagine that the child is presented with a task in which two arrays of elements (say, three sticks differing in size and three people differing in some attribute) must be mapped onto each other, by analogy, at each level of mapping complexity. The three levels of mapping are as follows:

Element mapping, where an individual element in one structure – for example, the element that represents 'a stick' – is mapped onto a single element of another structure – for example, the representation of a person named John, with no relations between the other elements within each structure constraining the mapping; any old stick will do as an elemental mapping for John.

Relational mapping, where each element in the *mapping from* structure (sticks) is compared with each other element, pair-wise, and this relationship is mapped onto the relationship between individual elements in the *mapped onto* structure (people). For example, the element mapping described above would become a relational mapping should the relations between the elements in the *mapping from* representational structure ('Stick A is bigger/taller/nicer than stick B') be preserved in the *mapped onto* structure ('John is bigger/taller/nicer than Mary'). No longer will any old stick do as a mapping for John if the people set is to maintain the relational mappings of the stick set. The simplest of analogies ('Doctor is to patient as Lawyer is to . . .') can be solved on the basis of relational mappings.

System mapping, where the mapping is from the systemic relation of one representational structure to another. An example is where the system properties of transitive inference are mapped between two structures. Thus, the representation of sticks as an ordered set of sizes may map onto an ordered set of representations of people whose relations are those of, for example, age. If stick 1 is bigger than stick 2, and stick 2 is bigger than stick 3, then stick 1 *must* be bigger than stick 3. For example, for an attribute like age, this system property would map onto: If person 1 is older than person 2 and person 2 is older than person 3, then person 1 must be older than person 3. In the case of a game in which a child pretends that sticks are people, if the child insists that the mapping from sticks onto imaginary people preserves the transitive relations of size in the set of sticks to ages in the set of imaginary people, then this would be an example of a system mapping. In system mapping a comparison between any two elements of one structure could generate the functional relations between any two elements of the other structure.

It is Halford's contention that this hierarchy of mappings imposes a concomitant increase in processing demands or loads. Thus, in an element mapping, only a single mapping must be computed between the elements of the different sets; in a relational mapping, the binary relation between each pair of elements within a set must also be mapped onto a corresponding binary relation in the other set; in a system mapping, at least two binary relations must be compared simultaneously. Halford argues that many tasks that are sensitive to the developmental changes that take place around 5 years old require a minimum of two binary relations to be considered simultaneously (including transitive inference, class systematicity in analogies, strategy use in beam-balancing problems, and class inclusion; see, for example, Halford and Leitch 1988). He proposes that the child below age 5 does not have sufficient capacity to consider more than one binary relation, and that this is the source of the difficulty for the child with these kinds of tasks.

In the next chapter we will have to consider the work of Case and Halford and the related studies of Kail (1986, 1988) in more detail. In that chapter we will look at the empirical claims advanced in support of the idea that processing speed, or capacity, not only changes with development, but is causally related to it. For the moment I will sum up this neo-Piagetian position from the perspective of the options available for development present in the minimal cognitive architecture and try to convince you that there are good theoretical grounds for rejecting the notion.

Implications for the Developmental Dimension

Both Case and Halford, like Piaget, see development as characterized by the growth in the ability of the child to solve increasingly complex problems. Unlike Piaget, however, they do not regard transition between complexity levels as constrained by the inherent shift in *logic* that they require. Rather, they view the increased ability of the child to solve increasingly complex tasks as a direct function of the increase of computational capacity available to them. For Case, this is brought about by the increase in efficiency of the simpler computations; for Halford, by an increase in the available capacity of the system. In either case the increase is in some kind of mechanism, or parameter of a mechanism, which is not an attribute of knowledge itself. That is, increasing efficiency, or capacity, is not caused by more efficient algorithms, or organization, but by a faster, or higher-capacity, neural architecture which implements the algorithms.

Such a view is equivalent in my theory to supposing that major changes in cognitive structure depend on increasing speed of the basic processing mechanism. Since major changes in development have been hypothesized to be caused, primarily, by the acquisition of new modules, this is tantamount to conceding that the operation of modules is dependent on variations in the speed of the basic processing mechanism. This, of course, violates a basic tenet of the individual differences theory. I pointed out at the beginning of the chapter that one of the constraints on the specification of the developmental option is that it must not transgress the axioms of the theory as it applies to individual differences. On these grounds alone, a strong version of increasing speed, or capacity, as the *cause* of developmental change must be rejected.

Remember, though, that the developmental specification acknowledges two kinds of development. Could the second kind, the elaboration of knowledge structures and the conceptual changes associated with it, not be motivated by increasing processing speed? Indeed, this could be so, but such a hypothesis seems unnecessary. Much of the evidence for the speed, or capacity, link with conceptual change stems from the fact that children of higher measured speed, or capacity, can solve more difficult problems. But we do not need to infer that processing speed changes during development to explain this. It is an inherent property of the architecture (indeed, an axiom of the whole theory) that children who have higher processing speed will, other things being equal, be able to solve more difficult problems. One of the most important of the other things that must be equal is chronological age. So, processing speed will indeed predict when conceptual changes related to knowledge elaboration will occur (roughly in line with mental age, as discussed at the end of the last section) *without* supposing that processing speed changes in development. But what about older children? Is it not true that they are faster processors, or have higher cognitive capacity? Indeed, older children process certain kinds of information faster and are cognitively more capable than younger children, but we would expect this on anybody's story of development. Conceptual change could cause increasing cognitive capacity, rather than increasing cognitive capacity causing conceptual change. Thus, if the acquisition of a processing module or, more prosaically, familiarity with materials leads to a reorganization of knowledge, then we would expect the apparent capacity of the system to increase. The conclusion must be that since the hypothesis of increasing speed, or capacity, is superfluous (it doesn't do any theoretical work that we couldn't accomplish without it) and adopting it may violate an axiom of the individual differences theory, then we should reject the hypothesis and adopt the position that the speed of the basic processing mechanism, although it constrains knowledge elaboration, is itself unchanging in development. This also explains why the mentally retarded are not cognitively equivalent to MA-matched groups with normal IQs, and has the attraction of throwing in the stability in individual differences for free.

So much for a theoretical appraisal. But what about the mountain of evidence that seems to suggest that, in fact, processing speed, or capacity, or efficiency, does increase with development? In particular, Halford cites evidence that measures of capacity *which control for the role of knowledge in task performance* show increases during development that coincide with major changes in knowledge. In the next chapter I will evaluate this evidence, and provide some counter-evidence of my own.

7

Processing Speed and Development:Some Experiments

I have proposed that the processes underlying individual differences in intelligence and the processes underlying the development of intelligence are quite different. Individual differences are caused, in the main, by variations in the speed of the *basic processing mechanism*. Development involves two processes: *modular change* and the *elaboration* of existing knowledge structures. These processes are themselves related to individual differences in intelligence in different ways. The appearance of modules is unrelated to individual differences in intelligence, because the operation of the modules is unconstrained by the speed of the basic processing mechanism. However, the elaboration of knowledge will depend, in part, on the speed of the basic processing mechanism, as any knowledge acquired by thinking must be.

The developmental dimension complicates the notion of 'intelligence', arguing that there are at least two senses of the term rather than just one. The essential idea, however, is simple. The major causal factors in individual differences and developmental change are different. This makes it possible to entertain the hypothesis that, at least in one sense, intelligence does not develop. What I mean by this is that the speed of the basic processing mechanism (being the principal determinant of individual differences) could be unchanging in development and we would still find developmental change. Fine in theory, but what about in practice? Is processing speed unchanging during development?

I should make two things clear at the outset. First, the theory of the minimal cognitive architecture is, in fact, neutral with respect to the hypothesis that processing speed is unchanging with development. That is, the theory can easily accommodate the alternative hypothesis: namely, that processing speed changes with development. This is important, because it means that the theory does not stand or fall by the truth of the 'no change' hypothesis. Why then advocate it? This brings me to my second point. The value of the hypothesis is that it highlights the theoretical distinctions between the mechanisms underlying individual differences and those underlying developmental change. For the minimal cognitive architecture, it is at least *possible* that processing speed is unchanging with development. Indeed, if the hypothesis turns out to be *true*, then this would rule out many current theories of cognitive development and, simultaneously, leave the minimal cognitive architecture as the only theory that could accommodate the data on individual differences in intelligence and their influence on cognitive development. For these reasons alone, it is worth taking the hypothesis seriously. Besides, I happen to believe that the 'no change' hypothesis *is* true.

To understand the array of data that I am about to present you with, and perhaps to lighten your own cognitive load, let me translate the specification of the developmental dimension into the metaphor I used at the end of chapter 5. This may make the options that I want to contrast clearer. But the warning given in chapter 5 still applies: if you are distracted by metaphors or if you are inclined to take them too seriously then skip the next section.

Anderson and Anderson Revisited

We saw in chapter 5 that the success of law firms depends, primarily, on the speed of the secretary (basic processing mechanism). It doesn't matter how good Anderson and Anderson (the specific processors) may be at their job if the secretary cannot convert their work into the firm's product – namely, briefs and contracts (knowledge). But all law firms buy in the services of specialists (modules) whose products (representations) are made available to the firm independently of the secretary's speed – because she doesn't have to type them! So a secretary, two partners, and specialists are the minimum requirement for a functioning law firm. Individual differences are determined principally by the speed of the secretary and, to a lesser degree, by the abilities of the two partners. So much for individual differences.

It is also striking that firms that have been established for longer produce more briefs and contracts and are more successful. Now the Law Society has handed out another research contract: why should older firms be more successful than younger ones?

At first there was some resistance to this research, since the answer to the question appeared obvious to many of the members. Firms that have been around longer have more contacts and an accumulating base of expertise, whereas firms just starting out have to build up this base. Mystery solved. However, a preliminary report (chapter 6) convinced many members that this was too simple, since it seemed clear that there are qualitative differences between older and younger firms. It is not just that the more established firms do more work, but they do more *complicated* work. Why should this be?

At the end of the preliminary report two reasons were proposed:

1   As they grow, firms acquire the services of more and more specialists. These specialists provide complex products that are unconstrained by the speed of the secretary. So, for example, the intricacies of company law (which are too complex to be handled by either of the partners, especially if they are expected to handle other work too) are dealt with by standard packages bought from these specialists. The acquisition of specialist services is the major reason why more established firms are more successful and competent than newer ones.

2   The limiting case is that of firms who buy these services and leave it at that. These firms grow enormously in competence, but are not successful compared with other firms established about the same time or with some others that are quite a bit younger. This is because partners can make use of these new specialist products in more or less intelligent ways, and some firms have partners and secretaries better able to exploit their services. So, for example, sections of completed contracts purchased from a specialist may be usable in new, but idiosyncratic, contracts. Imagine the following case: there is a dispute over the transfer of property rights between a divorcing couple, one of whom is a British subject, the other an American citizen, where the property in question is in Bermuda and where the grounds for divorce are the husband's abuse of the only heir to the couple's fortune. In such a case a firm will have to improvise to meet the particular needs of one client. An established firm may well have bought in the services of specialists dealing with (a) divorce, (b) property, (c) investment abroad, and (d) child abuse, and if a partner in the firm can extract the relevant sections, they can be pasted together along with some connecting clauses to generate a new, complex, idiosyncratic contract. The ability of partners to *paste together* these bought contracts will, of course, depend on their talent and on how fast the secretary can type. While the secretary does not have to re-type the sections cut out of the specialists' contracts (these can be transferred wholesale by her word processor), she must still type the new sections generated by one of the

partners. The partners can only make these sections as complex as her speed will allow. So, although firms grow in competence, there are still stable individual differences in success.

The Law Society was, in the main, convinced by this story. Then a strange thing happened. Some of the members argued that no one had considered a most obvious explanation for why older firms are more successful than younger ones: namely, that older firms had faster secretaries. I say this is a strange thing because, ironically, the members who pointed this out came from the same firms who were originally incredulous that the speed of the secretary explained most of the differences between firms! So is it the case that the cause of the differences between older and younger firms is the same as that which causes differences between firms generally, namely, the speed of the secretary? Or is it the case, as was argued above, that the differences are to be found in the acquisition of specialists and the steady accumulation and elaboration of expertise?

These are the options that we will evaluate in this new report (chapter):

1   As firms become more established, the speed of the secretary increases, allowing more voluminous and complex work to be done.
2   Alternatively, the speed of the secretary does not change with the development of the firm. What does change is the availability of specialist services and the accumulation of expertise with experience.

Evidence for Changing Processing Speed

Let me be clear about what would, and what would not, constitute evidence that the speed of the basic processing mechanism changes in development. First, what does *not* constitute evidence?

Evidence that older children complete cognitive tasks faster than younger children is not sufficient evidence for age changes in speed of processing. This is because increased problem-solving speed can come about through changes in the way in which problems are tackled, as well as through increases in speed of processing *per se*. We can see this by expressing the options in terms of a computer metaphor. There are at least two ways in which a computer can become faster at solving problems. The first is related to the quality of the *software*. Do the algorithms used in problem-solving become more efficient during development because, for example, they have access to more elaborate knowledge structures (Chi and Ceci 1987)? In this case we become quicker at solving problems with age because we do things differently, rather than do the same things faster. The second is related to the operating characteristics of the *hardware* which implements problem-solving algorithms. Do aspects of the architecture improve in such a way that the implementation of old algorithms becomes faster and/or new, more complex algorithms become possible?

Increases in the speed of the basic processing mechanism would come under the hardware category of change; yet increases in speed of problem-solving could be caused by software changes. Therefore, the increase in problem-solving speed during development does not necessarily imply an increase in the speed of processing.

The attempt to disentangle these two positions has been at the centre of a long-running dispute among developmental psychologists (see, for example, Chi 1977, Chi and Ceci 1987), and as yet the issue remains unresolved. However, there is a growing body of influential theorists who back the view that the capacity of the hardware changes during development. These theorists disagree about the precise locus of developmental changes: whether in processing speed (Case 1985; Kail 1986, 1988; Dempster 1985) or in resources, or capacity, available to the processing system (Halford 1985). But, for the purposes of this chapter they are equivalent, in that they propose that changes in hardware, rather than software, promote cognitive development.

So what would I accept as evidence that the speed of the basic processing mechanism increases during development? There are two conditions that must be met: first, it must be shown that a parameter of processing that is independent of knowledge (that is, the very kind of information processing that the notion of the basic processing mechanism was designed to capture) changes with development; second, this parameter of processing must correlate with *individual differences in intelligence*.

As we examine the data that purport to show that capacity, or speed, changes with development, we should keep in mind the second of the two conditions. The capacity, or speed, in question must be related to individual differences in intelligence, within age-groups, to qualify as an index of the speed of the basic processing mechanism. This turns out to be an important proviso.

Before we begin, let me offer fair warning. In the remainder of the chapter, we will be going into the kind of fine detail that may not captivate those without a vested interest in the outcome. If you are perfectly happy with the idea that processing speed may be unchanging during development, then you could skim-read the rest of the chapter. However, if you think the idea is quite preposterous because *everyone* knows that something like processing speed or capacity increases during development, then you should read these sections carefully and give me a chance to persuade you to change your mind.

Measures of M-capacity

There is no better paradigm for highlighting the problems inherent in attributing increasing problem-solving ability to increasing capacity, or speed, than the M-capacity paradigm of Pascual-Leone and Case. Although they differ in their detailed theoretical accounts of why capacity changes during development, the empirical base of the hypothesis is the same. Both claim to have measured the knowledge-independent processing capacity of children and to have shown that it increases during development.

The measurment of M-capacity essentially involves a task analysis of the number and nature of cognitive operations required for solution of a task. The next step is to estimate how many of these operations can be invoked simultaneously by children of different ages. So, for example, in the compound visual stimulus (CVS) task, children learn to associate an arbitrary response, say a handclap, with an arbitrary visual stimulus, say a triangle (see figure 7.1). Another response, say a thumbs-up sign, may be associated with a circle, another with a square, and so on. When the experimenter is sure that each of the individual associations has been overlearned, the child is then presented with a CVS: that is, a figure composed of a superimposed triangle and a circle. The child then has to produce each of the responses associated with each of the elements of the compound stimulus. The CVS can be varied in complexity simply by increasing the number of elements (schemes). The measure of M-capacity is the number of elements in the most complex CVS that a child can successfully 'process'. This ranges from one element at age 3, and increases by half an element a year, to about seven elements at age 16. No one disputes that this *measure* of M-capacity increases during development; but what are the grounds for believing that changing M-capacity reflects changing *knowledge-free* properties of processing?

The supporting argument consists of two parts. First, measures of M-capacity successfully predict performance on a variety of cognitive tasks that are apparently unrelated in cognitive content to either the CVS or each other. This suggests that M-capacity is a knowledge-independent parameter of information processing. The second part of

Figure 7.1 An example of a compound visual stimulus (CVS) task. A child is taught to clap his hands in response to a triangle and to give the 'thumbs up' to a circle. When presented with the compound stimulus consisting of a triangle and a circle, he must produce the learned response to each of the elements.

the supporting argument depends on the validity of the first. If M-capacity is itself independent of knowledge and measures of M-capacity increase during development, then increases in M-capacity must be a causal, rather than a dependent, variable in cognitive development. QED.

The reader may well recognize this line of argument, as it is also central to the arguments marshalled in favour of general intelligence. The first part of the argument above simply refers to a positive correlation between performance on different tasks. But, as we saw in chapter 3, a positive correlation between performance on tasks that differ in content is not sufficient in itself to establish a causal connection between knowledge-free processing and the intercorrelation of cognitive abilities. So the first part of the argument above does not constitute sufficient grounds to justify the belief that general intelligence is a knowledge-free property of information processing, and the same applies to the hypothesis that measures of M-capacity reflect knowledge-free processing. Independent evidence must be mustered to support the contention that the basis of the intercorrelations is processing capacity, just as independent evidence for a processing speed/general intelligence link had to be provided in chapter 3. Within the M-capacity paradigm, no such independent evidence is forthcoming, as we shall see. The crucial concern, then, is whether M-capacity is knowledge-free or whether changes in M-capacity could be caused by changes in knowledge.

There are a number of reasons for supposing that, rather than being knowledge-free, measures of M-capacity depend on knowledge-rich processing:

1   Arbitrary responses must be associated with arbitrary stimuli, a feature of the paradigm that takes time and training.
2   A compound stimulus must be analysed for its constituents and compared with the stimulus set in memory.
3   The set of responses must be searched for an appropriate match to the stimulus.
4   Responses must then be selected and executed.
5   Meanwhile, the original compound stimulus has been removed from view and must be 'imagined' afresh so that the next process can begin again.
6   Alternatively, the child may choose to try to decode all the stimuli and select all the responses before the 5-second presentation of the compound stimulus ends, after which the task becomes one of remembering a list of responses.

The point is that the measure of M-capacity involves tasks with abundant possibilities for strategic changes to give rise to large developmental differences in the measure. Recall that even Jensen's RT procedure was shown to have knowledge-related components that influenced performance and his procedure was very much simpler than the one used to assess M-capacity. Indeed, it may be *because* the task is sensitive to a myriad of strategic changes in processing that it is a good developmental measure. Further, Halford, the other major capacity theorist, argues that Case's procedure for measuring M-capacity leaves the notion of capacity empirically empty.

Remember that in Case's theory, development comes in four main stages, each of which has the same four substages (see chapter 6). Each of the substages is characterized by different *short-term storage* loads. These are measured in S units, where S is the number of pointers that must be maintained in a working memory system. Halford (1987) points out that while the short-term memory loads vary for each of the four substages (from 0S to 3S), when the child moves to a new stage, the processing load of a task is reset to zero. In other words, the capacity demands of the task are reduced. Since the measure of M-capacity depends on the operating schemes in use and these change from one stage to another, a change in operating scheme which is concomitant with a change in capacity is causally ambiguous. There is no way of telling whether increases in capacity give rise to changes in operating schemes (the hardware hypothesis) or whether changes in operating schemes give rise to increases in capacity (the software hypothesis). Halford claims to have solved this problem in his own work by both theoretically and empirically separating the capacity requirements of different cognitive structures from the resources or capacity available to the cognitive system at different stages of development.

Structure Mapping and Processing Capacity

The theoretical wing of Halford's position is the claim, as described in chapter 6, that different tasks can be analysed in terms of the *structure mapping* required for task solution. As the level of structure mapping inherent in the task increases, so does the number of binary relations that must be considered simultaneously, with concomitant increases in processing load. The increasing capacity demands of such mappings are independent of the cognitive system that happens to be doing the task; thus these mapping requirements impose the *same* processing

[1] An important plank in Halford's argument is the claim that structure mapping outlines the minimal mapping required by seminal kinds of reasoning processes that cannot be circumvented by strategic ploys. For example, in the case of transitive inference, no

loads on everyone.[1] This provides Halford with a true difficulty metric for different tasks, which can then be assessed for their processing capacity requirements.

The empirical wing of Halford's position is the advocacy of independent measures of capacity. He argues that a child's processing capacity should be measured using tasks that do not invoke the very same operating schemes which that capacity is hypothesized to constrain. This is the very thing that Case fails to do. Case notes that older children can process more stimulus elements than younger children in an M-capacity task. He claims that the reason for this is that the task, of course, measures capacity. But this is what is disputed. He cannot know that his M-capacity task measures capacity unless he has an independent estimate of this capacity and can show that it predicts performance in the M-capacity task. Halford argues that a dual-task procedure allows an independent assessment of a child's processing capacity. In such a procedure, the different capacity requirements of tasks are calculated by observing their effect on performance of a secondary task (usually reaction time). Tasks which cause greater performance decrements on the secondary task are considered to have greater capacity demands. In addition, Halford uses Hunt and Lansman's (1982) 'easy' to 'hard' procedure. This controls for the possibility that the estimate of an individual's capacity, calculated by measuring the influence of an 'easy' primary task on a secondary task, subsequently predicts performance on a third 'hard' task because of shared task features with the primary task used in conjunction with the secondary task (see figure 7.2). The easy-to-hard procedure statistically separates out the similarities between the tasks. This means that the ability of dual-task performance to explain

---

amount of 'chunking', say, can circumvent the need to make two relational comparisons, A > B and B > C, therefore A > C. To the extent that strategies *could* be invoked to circumvent the need to compare the two binary relations A > B and B > C simultaneously, then the subjects are not really doing transitive inference. So, 'transitivity entails a system mapping because of the definition of transitivity. We cannot write a new model of transitivity [for this read: the subject cannot adopt a strategy] that can be handled by a relational mapping because the new process would not be transitivity, *by definition*' (Halford 1987, 58; emphasis added). This is an ingenious argument, but specious (in this context) given his own critique of the empirical indeterminacy of Case's theory. Halford is not claiming that strategies cannot be invoked to solve transitive inference problems, just that if they are, this would not count as an instance of transitive inference. The problem is how we know whether a task that apparently requires the simultaneous comparison of two binary relations is actually being processed in that way by an individual child? We could always claim that a child who had less than the hypothesized processing capacity necessary for solving transitive inference problems but could nevertheless solve them was using a different strategy. This may leave Halford's conception of capacity in the same limbo as that of Case.

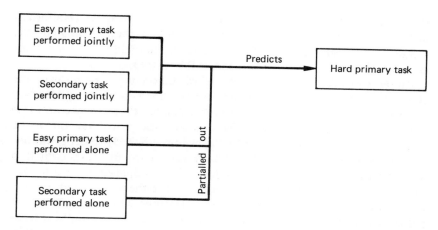

Figure 7.2 *The logic of the 'easy-to-hard' procedure. The correlations between the easy primary task and the secondary task, when performed alone, are partialled out of the correlations between the two tasks performed together (the dual task) and the hard (ex hypothesi, capacity-consuming) version of the primary task. This procedure removes any variation in performance of the hard primary task that may simply be due to shared characteristics of the easy and secondary tasks. The logic is that any correlation that remains between the dual task and the hard primary task cannot be due to common features of each task other than their capacity requirements.*

performance on a third task can be more surely attributed to the capacity requirements of the task. We shall examine an example of this procedure in more detail below.

From estimates using dual tasks, Halford found that a child's capacity increases with development. He then tied the two sides of his theory together, maintaining that cognitive development can be characterized as a progression through stages of increasing complexity of cognitive operations. What defines complexity is the level of structure mapping that a child can attain, and what accounts for this improvement is increasing processing capacity. The causal relationship cannot be the other way round, because capacity has been assessed on the basis of tasks for which the more complex structures are not invoked.

Halford's dual-task procedure and the resulting claim that capacity increases with development represent a substantial challenge to the unchanging processing speed hypothesis. Fortunately (for my hypothesis), a detailed examination of the methodology reveals that all is not as simple as at first appears.

A typical example of the empirical evidence mustered by Halford for the notion that increasing processing capacity is the motor of cognitive development is an experiment by him and his colleagues (Halford *et al.*1986). Using the easy-to-hard procedure, they estimated the influence of processing capacity on the ability of children to solve transitive inference problems of the sort A > B > C > D.

An 'easy' version of the task (where only adjacent pairs – for example B and C – are compared) is performed both alone and with a secondary task. The secondary task is also performed alone. Performance on these measures is used to predict performance on the 'hard' version of the primary task (transitive inference; for example B ? D) making the following assumptions: First, if the hard version of the task (which is always presented alone) is capacity-limited, then dual-task performance measures will be better predictors than will either 'easy' task performance or secondary task performance measures when these tasks are performed alone. Second, increased prediction is not simply due to similarities between the tasks. This possibility is controlled for by separating out the variation in hard-task performance predicted by both the 'easy' and the secondary tasks *when performed alone*. In this way the knowledge-based component that the 'hard' version of the task has in common with the 'easy' version is assumed to be separated out.

Halford *et al.* (1986) found that dual task performance significantly predicted the variation in transitive inference, even after the variation predicted by each task performed alone was removed. In addition, there was some tentative evidence in the study that processing capacity increases with age. This evidence seems to suggest that the ability to make transitive inferences is capacity-limited and that increasing ability with age is due to increasing processing capacity. But does it?

What is clear is that variation in subjects' ability to perform two tasks simultaneously predicts significantly more variation in the performance of the 'hard' task than does variation in performance on either of the tasks alone. But is this because the dual task is assessing spare capacity? Or could it be that the structure of each task interferes with the other? (If I have to juggle a ball with my hands, this is very much harder to do if I must simultaneously write my name, and I do not have to invoke the notion of capacity to explain why). Navon (1984) has used this counter-possibility to attack the whole concept of cognitive capacity. A closer look at the dual-task situation of Halford *et al.* makes this alternative hypothesis look plausible.

Halford *et al.* (1986) used children's judgements about remembered sizes of rods (for example, that the red rod is bigger than the green rod). The experiment is illustrated in figure 7.3. Five perspex rods, *a* to *e*, of decreasing size, were each kept in a coloured cigar tube of the same length. A memory board was always available for view. The

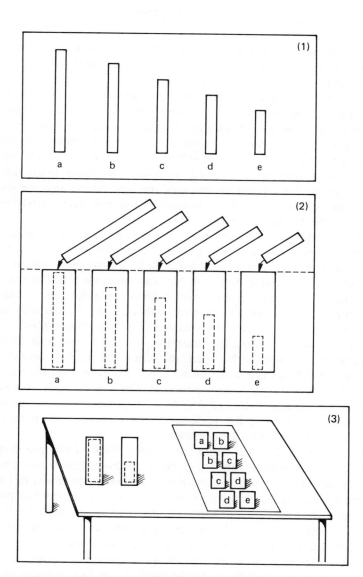

Figure 7.3 *Rods of different lengths (1) are put in different coloured tubes (2) which are all the same size. Children are taught which size rod is in which coloured tube and the size relationships between adjacent size pairs (for example, a and b, and b and c). The size relationships of adjacent pairs are also represented on a memory board, which is in view throughout the experiment (3), with the larger of any pair on the left. The child's task is to say which tube contains the largest rod when presented with non-adjacent pairs (which requires him or her to make a transitive inference). After Halford* et al. *1986.*

memory board contained four pairs of pegs which were the same colours as the cigar tubes. The pairs of pegs were arranged to represent the relative sizes of the rods contained in the cigar tubes. Say the longest rod, *a*, was in the yellow tube and the second longest, *b*, was in the red tube, then the yellow peg would be paired with the red peg, with the yellow on the left. In this way the pairings $a > b$, $b > c$, $c > d$, $d > e$ were represented as coloured pairs of pegs with the longest always being to the left of the pair. The 'hard' primary task (necessitating transitive inference) involved ordering two tubes (with the one containing the longer rod on the left, the one containing the shorter rod on the right) whose size relationship was not immediately represented on the memory board.

The first oddity in this experiment is the nature of the 'easy' primary task, which involved the ordering of two tubes that were represented on the memory board. The assumption that this is an easy version of the same kind of task seems doubtful. No reasoning, even of the easy variety, need be involved in performance: simply adopting the strategy of matching the presented colour pair with one on the memory board would guarantee successful performance. Contrast this with other transitive inference paradigms (for example, that of Bryant and Trabasso 1971) where the *relation* between adjacent pairs in the ordered array may be overlearned but must be recalled, ensuring that, for example, the relation A > B is, at least, *represented*. There is no such necessity in the Halford paradigm. Indeed, the 'easy' task was too easy when performed alone, with all the children's scores being perfect; therefore there was no variation in performance.[2]

This may seem a rather minor technicality. 'Well, OK,' you may say, trying to suppress your irritation at all this tedium, 'the "easy" primary task was too easy but not *so* easy that performance on it was not hit by pairing it with the secondary task; and, anyway, the substantive point is that the *secondary* task performance declined when paired with the "easy" primary task.' This brings me to the second oddity: namely, the nature of the secondary task. The secondary task was a 'span test' in which the children had to remember pairs of stimuli drawn on cards. How many cards could be correctly recalled was the measure of span. The problem is that the stimuli pairs were *colours*. So a span of 3 might involve remembering that red was paired with green, blue with yellow,

---

[2] Incidentally, because there was no variation in performance of the 'easy' task, partialling out the correlation of individual variation on the 'easy' primary task with that on the 'hard' task (to prevent any increased predictability of the dual-task performance being due to higher facility of some children with the 'easy' version of the task) is meaningless. So one control for the capacity hypothesis is removed.

and orange with pink, or whatever. It doesn't take a great deal of imagination to see how having the secondary task utilize the same stimulus dimension (colour) as the 'easy' primary task could produce a decrement in performance on each task when performed together that has nothing to do with the competition for a common resource. The dual task may test the child's ability to synchronize the separate tasks so that the stimulus encoding, decision making, and response output features of one task do not clash with that of the other. In Halford *et al.*'s study this must have been a bit like trying to juggle and write at the same time. There is considerable doubt, then, whether the procedure used by Halford and his colleagues assesses knowledge-free capacity or whether, on the contrary, it reflects differences in the ability to optimize knowledge-rich processing strategies.

Although Halford is sensitive to the problems of the processing capacity paradigm and has gone to great lengths to obviate them, it should be clear that the case for processing capacity (as a knowledge-free parameter of cognition) increasing during development is not proven.

Supporters of the 'no processing speed change with development' hypothesis may be starting to feel relieved; but unfortunately there is yet more trouble on the horizon, this time in the shape of processing speed itself and the claim by Kail (1986, 1988) that it increases during development.

Processing Speed

Kail (1986, 1988) claims that cognition is limited by a central processing mechanism that increases in its speed of operation during development. His evidence is that the function describing increasing processing speed with age is identical for two quite different processes. The speed of these processes was measured using two tasks: (1) a *mental rotation* task (Cooper and Shepard 1973) involving a decision as to whether letters, displayed in a variety of orientations, were identical or the mirror image of an upright letter presented simultaneously; (2) a *name-retrieval* task (Posner and Mitchell 1967) in which pictures of objects that were either physically identical (same picture) or had the same name (picture not identical but involving the same object) or were of different objects were presented. There were two different response conditions in which the identity of two stimuli had to be judged the same or different. In one, subjects responded 'Yes' if and only if the stimuli were physically identical; in the other, they responded 'Yes' if they had the same name.

Analysis of task performance revealed that the appropriate informa-tion-processing operations were being tapped and that these operations were conducted faster by older children.[3] The growth curves describing increasing processing speed with age suggested that (a) the increase in speed was the same for both tasks – that is, the rate of change param-eter for the two curves did not differ; and (b) the increase was due to improvements in a central limiting mechanism, because the develop-mental change in response time was best fitted by an exponential curve, rather than the hyperbolic curve typical of changes with learning or practice (Mazur and Hastie 1978).

In another study (Kail 1986, experiment 3), the hypothesis that changes in performance with age are due to a difference in a single lim-iting capacity, rather than a collection of specific improvements in a number of processes, was further tested using the mental rotation task. The logic of the experiment was to exploit the fact that the mental rota-tion task involves a number of processes which are affected by different experimental manipulations. Such manipulations created the different experimental conditions in the study. For example, changes in orienta-tion affect the process of rotation, and changes in stimulus quality affect stimulus encoding. The response times of a group of children and a group of adults were compared in each of 24 different conditions. The argument is that if all the processes involved in a mental rotation task are affected by a single limiting mechanism *and* if this mechanism increases in speed with age, then adults will be faster under all condi-tions, and the correlation between adult and children's response times *across all 24 conditions* should approximate 1.0. This is because the same increment in time is being added to all the processes, and correlating two scores that differ only by a constant produces a correlation of 1.0. On the other hand, if different processes are changing at different rates with age, then the scores under different experimental conditions will differ by more than a single constant, and the bigger the difference in rate of increase of the different constants, the more the correlation between adults' and children's response times across all conditions will deviate from unity. In fact, the reported correlation across all condi-tions between the two groups of subjects was 0.93, which, taking into account error of measurement, is effectively the unity predicted by a

---

[3] In the case of *mental rotation* it is the process of rotating a mental image in the mind which produces a linear relationship between response time and increasing orientation disparity between the stimuli. In the case of *name retrieval* it is the process of accessing name information from semantic memory: the physically identical stimuli were responded to faster than those which were physically different but of the same name, indicating the longer time it takes to retreive information from semantic memeory.

single change in a centrally limiting capacity.

Yet again, we seem to have convincing evidence that what changes in development is some knowledge-free parameter of processing. In the case of processing speed, of course, this is the very parameter I have argued is responsible for the majority of individual differences in cognition. Thus, the work of Kail represents the greatest challenge yet to the hypothesis that the speed of the basic processing mechanism is unchanging during development, and simultaneously threatens to undermine the thesis in chapter 6 that development and individual differences reflect different kinds of processes. But yet again, all is not what it might seem to be.

Stigler *et al.* (1988) have argued that a less restrictive version of a specific learning hypothesis, the skill-transfer hypothesis, is perfectly compatible with Kail's data. Skill transfer simply means that learning will transfer from one task to another to the extent that the two tasks involve the same kind of processing operations. They question the extent to which the mental rotation and name-retrieval tasks used by Kail involve distinct processes. Their argument is that if these tasks can be shown to have a number of common processing components, then the transfer of specific skill on any component from one task to another would produce the same data: namely, similar rate of change parameters with age for the two tasks. In support of this proposition, they cite a study by Kail (1987) himself, which shows extensive transfer of performance between mental rotation and name-retrieval tasks. They also point out that on the skill-transfer hypothesis we should expect that the component processes of a skill should develop at the same rate simply because, in all probability, they will be practised together. Again, the data reported by Kail cannot differentiate a central limitation from a skill- transfer hypothesis. Finally, Stigler *et al.* argue that a non-significant difference of only 3 per cent more explained variation, as found by Kail, is scant reason to prefer an exponential over a hyperbolic curve and to reject the hypothesis that developmental change can be accommodated by a standard learning model.

Kail (1988) retorted that there is plenty of evidence to suggest that mental rotation and name retrieval tap different abilities and that in any case the parameters in question (rotation rate and retrieval from semantic memory) were *derived* from task performance and are clearly different processes. For the same reason, Stigler *et al.*'s speculation about the simultaneous practice of component skills is spurious. On the significance of the superiority of an exponential over a hyperbolic curve, Kail points out that (a) it has been replicated; and (b) in training studies, learning curves are almost always hyperbolic and, therefore, *any* superiority of the exponential goes strongly against a learning hypothesis. This is a spirited defence; but there are other problems too.

In the crucial third experiment (Kail 1986), in which age changes in speed of mental rotation were investigated, the comparison groups did not differ only in age; they also differed in IQ. Thus, the older group consisted of 20-year-old undergraduates, whereas the younger group consisted of unselected 10-year-old schoolchildren. It is highly likely that the undergraduates were of higher IQ, which, *ex hypothesi*, means that they would be expected to have faster processing speeds. This is not a problem that Kail could have anticipated, but it means that the apparent age-related speed differences in this experiment may be IQ-related speed differences and hence that the experiment cannot be considered as counter-evidence against the hypothesis that the speed of the basic processing mechanism is constant with development.

In addition, Kail (1986, p. 984) himself points out that younger children are less accurate on mental rotation tasks, as well as slower, and that small age differences in accuracy can distort inferences about age differences in response times. The children were not less accurate on name retrieval, however. Perhaps they were trading off speed and accuracy in mental rotation and so, in order to achieve equal accuracy, they had to devote more processing time to the task. The consequence of this is that we do not know whether growth functions at equal levels of accuracy might yield significantly different growth functions for mental rotation and name retrieval.

In sum, the data cited by Case, Halford, and Kail do not present an unassailable case against the hypothesis that the speed of knowledge-free processing increases during development. Part of the problem (at least for Halford and Kail) resides in the complexity inherent in any reaction time task. The RT task, as we saw in chapter 3, leaves plenty of room for strategic differences to account for the developmental change usually cited as evidence for changes in processing speed. However, developmental changes have also been reported for *inspection time* (IT), suggesting that processing speed increases with age. I have argued that IT tasks are relatively free from strategic influences, at least compared with RT tasks. Clearly, this is worrying. If I criticize these data by claiming that changes in IT do not necessarily reflect changes in processing speed, this threatens to hoist my whole theoretical enterprise, painstakingly defended in chapter 3, by my own petard. However, claim this I will.

Inspection Time and Development

Most IT studies have investigated IQ-related differences in adult subjects, but some have looked at such differences in children (Hulme and Turnbull 1983, Nettelbeck and Young 1989). The absolute levels of IT reported in these studies seem to indicate that IT is longer for children than adults. However, only seven published studies have explicitly looked for developmental changes by comparing children of different ages (or children with adults) within the same study. The best of these is Nettelbeck and Wilson (1985).

Nettelbeck and Wilson investigated age differences for both cross-sectional and longitudinal samples, and also the effects of practice at different ages. In their third experiment they compared 10 children at each of 7 consecutive year levels, starting at about 7 and ending at 13, with a group of 10 first-year university students for measured IT. Those children who remained in the school for the next two years were retested twice, providing longitudinal data on changes in IT. Additional control groups of 10 7-year-olds and 10 12-year-olds, from a different school, and an additional 10 university students were tested twice, with two weeks between each session. This group acted as a control for *task-specific* practice. The principal findings were (1) that there was a significant trend of decreasing IT with age (although the most unambiguous data were for the 8 to 11-year-olds); (2) that practice had significant effects for the control groups, but that the effects did not interact with age; (3) that there were significant decreases in IT over the two one-year intervals between retesting, but that there was some evidence that 13-year-olds were approaching asymptotic performance levels; (4) that the size of the longitudinal effect was much greater than that of the practice effect, suggesting that the advantage that age brings is not task-specific, given that, if anything, the shorter time between retesting should have conferred the greatest degree of task-specific transfer in the practice control group.

Although it is true that the results are at odds with a previous study of children aged between 7 and 10 and adults, which reported no relationship between age and IT (Nettelbeck and Lally 1979), the study of Nettelbeck and Wilson can hardly be faulted. Noting the discrepancy between the two studies, Nettelbeck and Wilson (1985) could offer 'no interpretation for this discrepancy' (p. 20). Still, given the rigorous technique and sophisticated design *and* the important fact that the study was longitudinal, we have to believe that decreases in IT with age can be found. But does this mean that processing speed (as a knowledge-free parameter of processing) is similarly decreasing? If I still want to hold onto the hypothesis that processing speed does not change during development, I must be able to offer plausible explanations of why IT may change in development, for reasons other than that processing speed is changing. I must also bear in mind that these plausible explanations must not contradict the hypothesis that individual differences in IT *within* age-groups primarily reflect differences in the speed of the basic processing mechanism.

A reasonable hypothesis for why measures of IT may decrease during development while processing speed remains constant is that aspects of the task, incidental to processing speed, are more difficult for young children. Nettelbeck and Wilson recognized this possibility by changing the discriminative response from the standard *left/right* judgement to a spatial judgement involving *sides* (*this* side or *that* side). They did so because, as they themselves put it, the *cognitive load* of the concepts 'left' and 'right' may be higher for younger children. To substantiate the increasing processing speed hypothesis, we must rule out the possibility that younger children may perform more poorly on the standard IT task for reasons quite incidental to their ability to process stimulus information quickly – for example, IT could also decrease with age because older children are more attentive or more motivated, factors that may be important for performance on the standard IT task but which are unrelated to processing speed. What is needed is a task that measures IT which is more suitable for young children.

Space invaders

I developed a task to measure IT which was designed to reduce possible cognitive load differences between children and adults and remove some of the attentional and/or motivational differences that may underlie changes in IT with age. I therefore embedded an IT task in a computer-controlled video game (see Anderson 1986a for a full description).

In this task children must decide whether a space invader has two antennae of the same length or whether one is shorter than the other (see figure 7.4).[4] The invader is hidden by a 'bush', and is only revealed for short durations. The exposure duration can be varied, and, by measuring response accuracy, an estimate of IT can be obtained. During practice sessions children were allowed to specify their own discriminative responses (they could judge whether the invader had antennae that were 'the same' or 'different', or whether the invader was 'big and

[4] I want to thank Uta Frith for suggesting that the space invaders' antennae could be used for the line-length judgement.

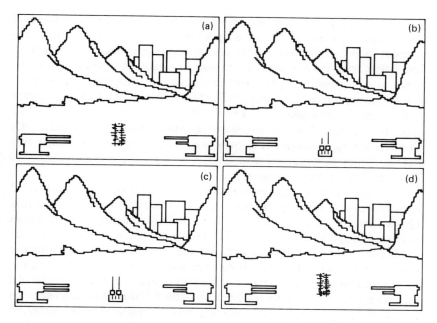

Figure 7.4 *In this test, children are told that a space invader is hiding behind the bush in the centre of the screen (a). When they push a button, the invader is revealed as in (b) or (c). The children have to push one button if the invader's antennae are of different lengths, as in (b), and another button if they are the same, as in (c). They do not know which they will be presented with in advance. The invader is exposed for a pre-set time, after which the bush reappears (d). The bush acts as a 'mask' for the stimulus (the space invader), destroying any information remaining on the child's retina, so the necessary information can be extracted only from the stimulus during the time it is shown on the screen, although the children can take as long as they like to make the response. Those with higher IQs have higher accuracies at shorter durations of exposure. Taken from M. Anderson 1990.*

little', as opposed to 'big and big', or whatever suited them) which were then associated with two different colour response keys. In addition, if a mistake was made during testing, then a 'dummy' non-scoring trial was inserted to avoid the ubiquitous repetition and alternation response strategies, following an error, from distorting the staircase algorithm. Other features were added, such as feedback for every response, rewards of whooping spacecraft taking off after a number of trials, and so on, all with the aim of removing non-processing speed influences on task performance and motivating the younger children to attend. With such a task, I believed, apparent differences in processing speed would disappear.

Is the IT measured in such a task more related to age or to individual differences in intelligence during development? To the extent that differences in IQ reflect differences in processing speed *and* to the extent that differences in IT reflect differences in processing speed *and* to the extent that processing speed remains constant through development, *then* IQ (irrespective of the age of the child) should be a better predictor of IT than chronological age. On the other hand, if processing speed changes with age, then CA should be a better predictor of IT than IQ, which is an age-standardized measure. The experiments that I will describe below investigated whether CA or IQ is the better predictor of IT in children of different ages.

In a series of three experiments (Anderson 1986a) IT was estimated in 44 children in three age-groups of 6, 8, and 10. All the children had IQs within the normal range. There was no significant difference between the age groups in mean IT. However, when the children were divided up into low- and high-IQ groups (irrespective of age) on the basis of their full WISC-R intelligence test scores, the low-IQ group were found to have significantly longer ITs. Further, the mean IT for high-IQ 6-year-olds (mean IQ 121) was 156 milliseconds, whereas that for low-IQ 10-year-olds (mean IQ 97) was 214 milliseconds. Although the difference was not statistically significant, it does suggest that increasing age cannot compensate for lower IQ.

The foregoing paragraph is a little disingenuous, as it obscures a complication in the data. While it is true that, in general, IQ was a better predictor of IT than CA in these studies, IQ was found to be not as good a predictor as mental age (MA). We perhaps need to recap on these psychometric distinctions to understand what this might mean.

Both IQ and MA can be derived from a child's score on an intelligence test, for example, Raven's Progressive Matrices. An IQ is calculated by comparing the *raw score* (that is, the actual score obtained on the test) with the distribution of scores on the test for children of the same age (age-normed). The raw score can also be used directly as an estimate of mental age – that is, where the scores are treated as absolute values on one continuous scale of cognitive achievement. Although developmentalists tend to think of cognitive development as being chronological age (CA)-dependent, psychometricians view the development of intelligence as being indexed by MA, *not* CA. Thus, psychometric theories, which view intelligence as a reflection of processing speed and hypothesize that increasing processing speed underlies the development of intelligence, predict that IT should be more related to MA than CA. The finding that IT relates to MA more than IQ in development would seem to support their hypothesis and undermine my own hypothesis that processing speed doesn't change during *cognitive* development. But what is MA from the perspective of my theory?

Remember that there are two developmental processes. Modular change is the primary developmental process, and is unconstrained by differences in the speed, or efficiency, of the basic processing mechanism. However, knowledge *elaboration*, the second kind of developmental process, will be constrained by speed of processing. We can assume that increasing MA reflects both these processes and that because of the relationship between processing speed and knowledge elaboration, IQ and MA will be correlated. MA will be expected to pick up some of the IQ/IT common variance (that due to a shared reliance on processing speed) and also any additional variation in cases where increasing knowledge stands to aid task performance. In effect, MA can capitalize on the best of both worlds, in that it will predict the contribution that variations in both processing speed and knowledge make to performance on IT tasks. This leads to a clear prediction: that the relative contributions of IQ and MA to task performance will vary with the extent to which the task is knowledge-free. The more knowledge-free it is (given the hypothesis of unchanging processing speed), the greater the correlation with IQ. The less knowledge-free it is, the greater the correlation with MA. This hypothesis was tested in a subsequent experiment (Anderson 1988).

On the assumption that tasks can be arranged on a continuum that runs from knowledge-free to knowledge-rich, it is clear that IT is more knowledge-free than RT (see chapter 3). The correlations with IQ, MA, and CA were compared for an IT task and an RT task that tested the retrieval of information from semantic memory (Posner and Mitchell 1967), a process known to undergo developmental change (Keating and Bobbitt 1978). This RT task was one of those used by Kail (1986) to illustrate the development of processing speed and is described in the section on Kail above. The relevant correlations were computed for 93 children, ranging in age from 8 to 12, and the results were clear (see figure 7.5).

IT was still found to relate more to MA than to IQ, but the fact that it relates at all to IQ, given such a wide age range, is remarkable, as evinced by the fact that there is absolutely no correlation between RT and IQ. RT, in fact, is highly related to CA. In other words, IT and RT are differentially related to age and individual differences, and in a way that is broadly consistent with the hypothesis that processing speed is responsible for differences in intelligence but is largely unchanging throughout development. What is changing is knowledge, and changes are better detected by a RT index.

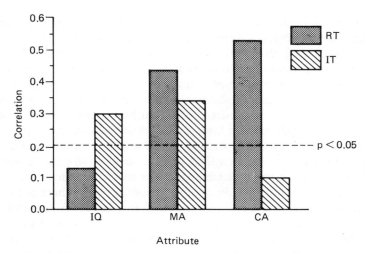

Figure 7.5 *The data, taken from Anderson 1988, are for 93 children aged between 8 and 12. IT was derived from a method of limits procedure. RT refers to the standard deviation of RT (since it is a better correlate of intelligence than mean RT) in a naming task.*

So it seems as if there is some support for the notion that processing speed doesn't change during development. IT is still predicted by individual differences in intelligence over a substantial age range. But if this is so, why are there clear decreases in measured IT with age? If there are developmental changes in processes that are normally unrelated to individual differences in processing speed but which contribute some variation to IT, then IT could decrease without any need to postulate changing processing speed. Two possible processes, attention and response selection, have been tested for their relatedness to age differences.

In a typical IT task the stimulus is presented for such brief durations that it is obviously critical that the child attends to the stimulus at the moment of presentation. There is some evidence that attentional factors may play a role in IT performance, particularly for children (Nettelbeck and Young 1989). In an experiment involving 40 12-year-olds and 32 8-year-olds, I investigated the effect of voluntary and involuntary attentional factors on age differences in IT (Anderson 1989a). In fact, although the usual age difference in IT was found in the study and although IT did vary with attention, there was no relation with age. Thus, although attention affects IT performance, age differences in IT are not due to differences in attention.

In another experiment I contrasted the effect of increasing the visual processing demands of the task with the effect of increasing response selection demands to see whether either of these influenced the age difference in IT (Anderson 1989b). I did this by changing the space invader stimulus so that it could have either four antennae on its head or only the usual two. The response requirements were varied by having two response conditions. In one condition the children simply had to report whether the space invader had antennae all of the *same* length or had one which was *different*. In another condition the response demands were increased by having them identify each stimulus separately (see figure 7.6). In other conditions the stimulus processing demands were varied by varying the number of stimuli that were presented.[5] The hypothesis was that if the difference between the age-groups were primarily a function of how fast the stimulus is encoded and if young children process more slowly, then increasing the stimulus processing demands should affect the performance of younger children more. If the difference is something to do with how reliably the subjects can choose among available responses, then increasing the response demands should affect the younger children more. In fact, the latter happened.

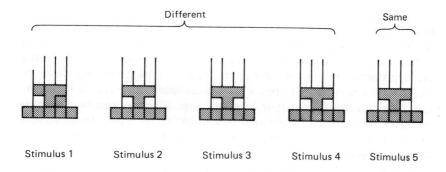

Figure 7.6 *The five stimuli used in Anderson 1989b can be classified as either 'same' (stimulus 5) or 'different' (stimuli 1–4), necessitating only two responses, or each stimulus may require individual identification, necessitating five responses.*

When the task was presented to 16 12-year-olds and 16 8-year-olds, increasing the stimulus processing demands increased all children's ITs, but did so equally for older and younger children. However, increasing the response selection demands increased IT in younger

[5] Note that as the response demands increased, so too did the stimulus processing demands. But suitable comparison of the ITs in different conditions allowed these effects to be separated.

children much more than in older children. This suggests that some of the age differences found in IT occur because older children can cope better with more complex response selection demands.

Could the ability to select a chosen response from a number of possible responses be modularly based? It would be unlikely to be based on a Mark I module, which is a specialized, computationally complex mechanism. However, such a competence could be based on a Mark II module, a component of the fetch-and-carry operations of information processing. For example, perhaps it is a developmental component of a response inhibiting system or, more prosaically, a cognitive skill developed through extensive practice. In either case, a developmental change unrelated to processing speed could influence estimates of IT. Therefore, decreasing IT with age does not *necessarily* imply decreasing processing speed. As long as variations in the developmental process are independent of individual differences in intelligence within subjects of the same age, then the conditions of supporting the unchanging speed hypothesis are maintained.

Conclusions

In this chapter I have not attempted to review all the literature on speed, or capacity, changes in development. Rather, I have challenged some of the strongest and most relevant studies that appear to falsify the hypothesis that processing speed doesn't change during development. At the very least, I hope I have established that there is considerable room for doubt that they have done so. What we can be sure of is that there are substantial changes in knowledge with development, and it is difficult to establish that there is anything left over for general changes in processing speed to explain. For example, Wickens (1974) concluded that 'Non-processing variables – practice, motivation, incentive, attentiveness – clearly affect children's maximum information-processing rate in experimental . . . situations to a greater degree than they influence the rate of adults.' And again: 'Although there appear to be strong differential effects of age on the latency of stimulus perceptual processing . . . the search data suggest that these do not reflect a central limitation but merely a learning or practice effect that naturally covaries with age. The data from tachistoscopic recognition studies are not conclusive with respect to this interpretation' (p. 752). It is clear that the question of speed, or capacity, changes in development is still a live issue fifteen years on.

Of course, faulting individual experiments and suggesting that there could be alternative explanations for the results obtained is often only a destructive exercize. However, in this case I feel justified in taking a strong sceptical stance. As I pointed out at the outset, if the 'no change' hypothesis turns out to be false, this would be easily accommodated by the current theory of the minimal cognitive architecture. But I have been careful to construct an *in principle* argument that processing speed may be unchanging in development, because this hypothesis gives so much theoretical purchase on the relationship between intelligence and development. Given this, much more pertinent contradictory data are required to justify abandoning the hypothesis at this embryonic stage. At the very least, we must wait for evidence that has taken into account my alternative hypothesis and has instituted the appropriate controls.

The theoretical purchase that the hypothesis gives is to emphasize the distinction between the processes underlying individual differences in intelligence and those underlying cognitive development. If the hypothesis is true, then those theories that attribute cognitive change to global changes in capacity, or speed, are wrong. If the 'no change' hypothesis turns out to be false, the distinction between the processes still stands, for the other reasons given in chapter 6.

In addition, if the hypothesis is true, then the explanation for one of the major regularities listed in the agenda for the theory (the stability of individual differences through development) comes free. If variations in the speed of the basic processing mechanism are the primary cause of individual differences in intelligence, then if this speed is unchanging, it is no wonder that individual differences show the stability they do. Changing processing speed could also accommodate this fact, but at the price of an additional assumption: namely, that processing speed increases are relatively constant across individuals.

In any case, most of the 'speed theorists' have it that changes in processing speed are most significant in the first five years of life (Case and Halford), and even the psychometrically minded acknowledge that any improvement in speed of processing is finished by about 12 years old (Nettelbeck 1987). Since there is still a great deal of improvement in psychometric test performance after 12, which must be caused by something other than changes in speed of processing, it is at least plausible that non-speed factors may also change cognitive performance before 12.

Adopting the hypothesis that processing speed is unchanging during development serves the purpose of highlighting the important theoretical distinction between intelligence and knowledge, on the one hand, and individual differences and development, on the other. It is clear that large-scale changes in knowledge are brought about not only by the maturation of *modules* (see chapter 6) but by the process of education that every child engages in as he or she grows up. And I don't just mean formal education, although it has been shown that there are large schooling effects even on so-called psychometric fluid abilities (Cahan and Cohen 1989); I mean all that happens to someone's knowledge as information is acquired and theories about the world are formulated. By teasing apart these different facets of intelligence, we can analyse their relative influence on cognition. To see what kinds of increased purchase this brings to bear on the relationship between intelligence and development, I will now turn, finally, to a re-examination of learning disabilities in the light of the new theory.

8

Patterns of Abilities: Regularities or Chaos?

Chapters 3 – 6 synthesized a great deal of evidence about the nature of individual differences in intelligence and their relationship to cognitive development. The result of this synthesis is the theory of the minimal cognitive architecture. Recasting old and often incompatible views and data is useful, but the real test of a new theory is whether it can generate *new* ideas and insights. We have just seen, in chapters 6 and 7, that the application of the theory to current developmental models has led to some new hypotheses about the relationship between cognitive development and individual differences in intelligence:

1   There are two developmental processes, the first being *modular change*. Modular change, determined as it is by mechanisms that are unconstrained by the basic processing mechanism (route 2 knowledge acquisition), will be related to chronological age, but will be independent of individual differences in intelligence.
2   The second developmental process is *knowledge elaboration*. Because it depends on the mechanisms that are constrained by the basic processing mechanism (route 1 knowledge acquisition), knowledge elaboration will be intimately related to individual differences in intelligence.
3   The speed of the basic processing mechanism is unchanging during development. This means that the development of, and individual differences in, intelligence *cannot* be different reflections of the same process. The constancy of processing speed also explains why individual differences are stable over age. Further, increasing MA with development cannot be due to changing processing speed, but is a cocktail of both modular change and knowledge elaboration.

I now want to try to use the theory of the minimal cognitive architecture to illuminate patterns of abilities: mysterious patterns that we already know about, but have trouble accommodating with other theories of intelligence. Moreover, we can begin to look for other patterns predicted from the new theory.

In chapter 3 I argued that the major causal variable underlying individual differences in intelligence was variation in the speed of the basic processing mechanism, and that this was the fount of general intelligence. I embellished this argument in chapter 5 by introducing the notion of specific processors and arguing that as processing speed increases, so the constraint imposed on the specific processors by the basic processing mechanism decreases. The major consequences of the theory for understanding patterns of ability are:

1   Low intelligence (in the individual differences sense of the term) is primarily the result of a slow basic processing mechanism.
2   Differences between specific abilities will be more obvious at higher levels of intelligence.
3   Individuals with a slow basic processing mechanism will be incapable of achieving the complexity of thought possible for individuals with fast basic processing mechanisms. In other words, for individuals of low intelligence some thoughts are unthinkable.

At least one well-known phenomenon seems at odds with these predictions: the isolated abilities of the *idiot savant*.

Anomalous Abilities

Idiots savants are individuals of low psychometric intelligence who can, nevertheless, perform remarkable cognitive feats. For example, one blind man with cerebral palsy who has an IQ of 40 and no spontaneous speech was observed, at the age of 12, to be playing Tchaikovsky's First Piano Concerto, which he had heard for the first time the previous evening when it was performed on television (Hermelin *et al.* 1989). O'Connor and Hermelin (1991) have studied another young man, with an IQ of about 75, who can translate texts written in German, French, and Spanish. During testing it became apparent that he could answer simple questions in Russian, Hindi, and Greek, and that if his IQ were to be based solely on his vocabulary, his English, German, French and Spanish IQs would be 121, 114, 110, and 89 respectively! Howe and Smith (1988) asked a 14-year-old-boy with an IQ of about 50 which the odd one out was of the seven months January 1971, September 1972, June 1973, July 1974, February 1971, August 1975, and October

1976, and were told (correctly) 'February 1971' because, unlike the others, this month began on a Monday (the others began on a Friday). Does the very existence of such people not belie the most fundamental axiom of the theory, namely, that the basic processing mechanism constrains *all* thought? For if these individuals are of low intelligence – that is, have slow basic processing mechanisms – how can they possess such amazing abilities? Indeed, for some, *idiots savants* call into question the whole reality of general intelligence (Howe 1990). Alternatively, if they are not unintelligent – that is, they have normal levels of processing speed – then why do they have low IQs, and why do these talents stand in such sharp relief to their other cognitive abilities? I hope to show that, rather than *idiots savants* being at odds with the minimal cognitive architecture, the pattern of *savant* abilities is made intelligible by the theory. In turn, this will highlight the advantage of having a theoretical specification of intelligence. Defining what intelligence is in precise theoretical terms allows us to solve the conundrum of whether such individuals are or are not 'intelligent'.

Who are Idiots Savants?

There are two defining characteristics of the *idiot savant*. Aside from showing an amazing aptitude for a single, often bizarre task, the most obvious characteristic is low intellectual ability. Both their everyday behaviour and their social interaction are characteristic of the mentally retarded. More often than not they cannot read, write, or count. It is important to note that even many of the calendrical calculators cannot reliably solve simple arithmetic problems. These behavioural indications of low intelligence are substantiated psychometrically by their low IQs. However, although low IQ is the norm, many display irregular intelligence test profiles (as of course do very many more 'normal' individuals), occasionally achieving near-average levels of performance on one or more subtests. For example, an artist *savant* who is in the bottom 1 per cent on the majority of thirteen subtests of the British Ability Scales administered to him edges into the bottom 15 per cent for two Block Design subtests, tests which are thought to measure spatial imagery, and is slightly above average on delayed visual recall (fifty-fourth percentile).[1] Obviously, relatively good performance on these subtests is consistent with his *savant* talent. Such cases suggest that the anomalous abilities of the *idiot savant* may be underlain by cognitive mechanisms that have been spared from the generalized brain damage

---

[1] Anthony Fallone, personal communication.

that we can assume caused their mental retardation. Such sparing can, on occasion, be detected by standardized psychometric tests. However, it must be said that their performance on even isolated subtests rarely, if ever, matches the degree to which their *savant* ability appears to outstrip 'normal' performance.

The best estimate of the prevalence of *savant* abilities in retarded people is about 0.06 per cent of the retarded population (Hill 1978). *Savant* abilities are relatively much more common in autistic populations, with the rate of incidence perhaps as high as 10 per cent (Rimland 1978). However, since mental retardation is far more prevalent than autism, only about a third of *idiots savants* will be autistic (O'Connor 1989). One feature of *idiots savants* is that they are predominantly male. Males are over-represented in mentally retarded populations generally, and are also more likely to be autistic. Both factors only partially explain why there is such an inflated male/female ratio, 6:1, for *savant* abilities (Hill 1978). O'Connor (1989), with reference to a sample of 60 *idiots savants*, estimates a 50 per cent prevalence of communication problems, finds a marked degree of obsessional preoccupation, and considers that the *savant* ability appears to have arisen 'unbidden, apparently untrained and at the age of somewhere between 5 and 8 years of age' (p. 4). It is likely, then, that there are a number of factors, including social awkwardness, an obsessional personality, brain damage, and an innate cognitive talent, that contribute to the phenomenon of the *idiot savant*. But let us return to the cognitive story.

The cognitive nature of Idiots Savants

It is striking that the *idiots savants* do not do just any old thing. *Savant* abilities cluster in six main areas: calculation, art, music, language, spatial skills, and memory (Sloboda *et al.* 1985). This rules out what I think of as a 'Martian virus' theory of the *idiot savant*. Such a theory would claim that the *idiot savant* has caught a rare virus that has come in from Mars on a meteorite. This virus radically rewires the brain of any individual who catches it, and in a way that is completely random with respect to psychological functioning. Most of the unfortunate victims die, of course, and the few that survive are distinctly odd. The point of this fanciful story is that on such a theory there are no implications whatsoever for human psychology of behaviour that depends on what is, in effect, an alien brain. This is why it is interesting that *idiot savant* abilities cluster in reasonably well-defined areas. For this seems to me to argue that the rewiring of the brain is not random. Why should random rewiring produce clusters of behaviour constituting recognizable, although bizarre, human abilities unless something about human psychology has either been preserved or has constrained the effects of the rewiring? This is why it is noteworthy that the same *savant* abilities occur in a number of different individuals. In the case of calendrical calculators, around twenty are known in the United Kingdom alone. It is extremely unlikely that random rewiring should repeatedly result in similar bizarre behaviour.

There is a more subtle variety of the Martian virus theory: Martian virus II. It could be that the domains of *idiot savant* abilities owe nothing to constraints of human cognition, but are demarcated on cultural grounds. That is, only activities that are recognized by the culture as embodying talent would be target *savant* abilities, and this explains why it *appears* that the *idiots savants* do not do just any old thing. If rewiring caused them to do something crazy (that is, any old thing), then it would pass unrecognized as an *idiot savant* talent. It is only when the rewiring happens to generate behaviour that we recognize as embodying talent that it will be so labelled.[2] This argument appears plausible for abilities in the fields of music and art, but is surely implausible for calendrical calculating (the talent manifested in about a third of all documented cases). This is an extremely obscure activity, and if one obscure ability can be spotted, it is likely that others would be too. It is not the case, then, that all *savant* abilities are acknowledged and culturally valued talents; some are clearly bizarre, and appreciated initially only for their curiosity value.

Contra either of the Martian virus theories, I believe that *idiots savants* can tell us something about the nature of cognitive abilities. What constrains the behavioural consequences of the brain damage suffered by *idiots savants* is their pre-morbid cognitive architecture.

I argue that *idiot savant* talents map onto the distinctions between the knowledge-acquisition mechanisms specified by the minimal cognitive architecture. They represent cases in which generalized brain damage has selectively spared just one of three crucial mechanisms: a module, a specific processor of high latent ability, or, more rarely, the basic processing mechanism itself. Most probably, in the majority of *idiots savants,* who are, we must assume, genuinely unintelligent (that is, have a slow basic processing mechanism), what has been preserved is a module. This must be so, because only modular functions are unconstrained by a slow basic processing mechanism. This hypothesis also makes sense in the light of the argument advanced in chapter 4 that

---

[2] In this case, of course, there would be only a superficial resemblance between the talent of an *idiot savant* and the same talent expressed in a normal person. This would render the *idiot savant* uninteresting for the study of human cognition.

modules are likely to be hard-wired processing mechanisms. Just as modules are prime candidates for specific damage, so too they are prime candidates for specific sparing of function, particularly an *isolated* function. I will now try to substantiate the hypothesis of a modular basis for *idiot savant* talents with reference to a particular case.

Language modules and linguistic savants?

Cromer (in press) has studied a case of a young woman suffering from spina bifida (with internal hydrocephaly) and displaying features of what has become known as 'cocktail party syndrome' (Hadenius *et al.*1962). This young woman (D.H.), in her late teens at the time of testing, has a very low measured IQ (verbal IQ 57, non-verbal IQ unscorable) with obvious cognitive consequences. For example, she cannot read, write, or count (despite intensive efforts at instruction), and cannot reliably order days of the week or seasons of the year. Yet she has remarkable linguistic skills. Her conversational language is fluent, pragmatically appropriate, and has complex syntactic forms with an extensive vocabulary and correct use of semantic constraints. Interestingly, her performance on various tests of language is nothing like as good as her spoken language would predict. Her performance on the Northwestern Syntax Screening Test (Lee 1971) and on a Test of Reception of Grammar (Bishop 1982) was poorer than would be expected even of someone of her IQ. She did better on the British Picture Vocabulary Scale (MA equivalent of 9 years, 5 months, IQ equivalent of 61). Yet, in many ways, her language use is indistinguishable from that of other young women her age; her fluent speech contains many grammatical features of the skilled language-user, such as relative clauses, passives, and complementary clauses. Nor is it the case that she simply performs poorly on all tests, for, on a hierarchical drawing test, a test that some aphasic children find impossible, she performs normally.

The conundrum of excellent natural speech but poor language test performance was solved by Cromer when he took the perspective that language tests do not typically 'measure those aspects of language that may be modular in the deeper and more modern linguistic sense'. Thus, when tested for linguistic *competence* by being asked to say whether sentences (spoken by a 'friend' of the experimenters, who did 'not yet speak English well') were acceptable or unacceptable English, D.H. made correct judgements regarding the use of prepositions, singular/plural agreement, tense markers, passives, and relative clauses. She did make mistakes on double-object constructions (for example, 'The Queen gave the school the toys', 'The man donated the library the book'; where the first is grammatically correct, the second not so), but then so do many non-retarded people. So it seems that, when formally tested, the things D.H. 'knows' about language are things to do with acceptable syntactic structures. On the other hand, her knowledge and understanding of the relationship between an utterance and some referent in the world (for example, judging whether a statement applies to a picture of a specific situation, a standard format for many language tests) is no better than what we might expect from someone of her low intelligence. We are thus forced to conclude that, although D.H.'s speech shows *linguistic* creativity and high performance, her *understanding* of what she says is characteristic of mentally retarded performance. In other words, were she to talk to herself, she would not be able to understand most of what she says. Note that this dissociation between general cognitive ability and linguistic skills exhibited by D.H. is by no means unique (Thal *et al.* 1989). The question is: Are we to believe that someone with these linguistic capabilities is unintelligent? An explanation based on the minimal cognitive architecture at least makes such a question intelligible.

In 1987 Richard Cromer and I tested D.H.'s inspection time, using the space invader game task described in chapter 7. D.H. could understand the task, and at long stimulus exposures (greater than 500 milliseconds) had no trouble discriminating the stimuli. But she was unable to make discriminations for exposure durations of less than 400 milliseconds, and refused to even 'guess', simply saying that she could not see the stimuli. I estimated her IT to be around 420 milliseconds, which is well into the mentally subnormal range. It seems, therefore, that her basic processing mechanism is inordinately slow, or inefficient. Since, according to the minimal cognitive architecture theory of intelligence, a slow basic processing mechanism is the primary cause of mental retardation (Anderson 1986b), she would be considered, on this theory, to be genuinely unintelligent. The fact that her isolated ability is linguistic is consistent with the idea that much linguistic knowledge is provided by modules whose operation is unconstrained by the basic processing mechanism. However, it is not clear why, in this case, her modular ability to process language has flowered into elaborate conversational skills rarely seen in other people of her mental age.

The hypothesis that *idiot savant* abilities have a modular basis is consistent with the fact that these abilities appear in isolation within a single individual. While it is true that some *idiots savants* have prodigious talent for more than one activity, these activities seem to be within-category variants. So, for example, the twins studied by Horowitz (see below) were not only calendrical calculators, but could also recognize prime numbers. However, as far as I am aware, there are no examples of *idiot savant* musicians who are artists or artists who are musicians. Nor are there linguistic *savants* like D.H. who also have musical and artistic talent. In true modular fashion, the ability is confined to one thing.[3] However, the hypothesis that *idiot savant* abilities are modular in origin is not without its problems.

Savant talents, such as linguistic, musical, and graphic art talents, are more plausibly module-based, because they seem to require distinctive and complex representational systems. However, given the strong hypothesis that modules do not show individual variation, or at least variation that is correlated with individual differences in intelligence, it is a little disquieting that musical abilities seem paradigm examples of *variable* talents (there are great, good, and not so good musical talents, and so forth) and also that they seem to depend on having a higher IQ than is normal (O'Connor and Hermelin 1983). Further, there are clear indications that modules cannot be the only source of *savant* abilities, as we shall see in the case of calendrical calculators.

Calendrical calculators

Idiots savants who have an amazing knowledge of dates and calendars are known as 'calendrical calculators'. They can answer questions like 'What day was 24 July 1955?' or 'What day will 13 October 2010 be?' or 'Tell me the date of the third Friday in May 1902?' or 'In what years will 19 December be a Wednesday?' The most celebrated examples are the identical twins Charles and George, studied mainly by Horowitz and subsequently popularized by Sacks (1985). The twins could answer calendar questions that spanned dates over thousands of years. There have been a variety of hypotheses put forward as to how *idiots savants* do their calculations:

(1) The least likely hypothesis is that *idiots savants* use one of the published algorithms for working out the calendar. For example, by following a prescribed number of steps, such as taking the last two digits of the year, dividing them by four and ignoring the remainder, adding on code numbers corresponding to months, and so on, any day of the week can be calculated for any date given. It is unlikely that *idiots*

---

[3] The only exception I know of is a musical prodigy who was also a calendrical calculator (Scheerer *et al.* 1945). However, calculators may, in any case, be a heterogenous group, as we shall see below.

savants use this method because the component operations, such as adding together two numbers, cannot be performed reliably by most calendrical calculators. Further, such methods are of little use for answering questions of the 'What years will the 24th of July be a Sunday' variety.

(2) The most favoured hypothesis is that these feats reflect an extensive rote memory for calendars (Horowitz *et al.* 1965, Hill 1978). Evidence for this hypothesis rests mainly on the fact that calendrical calculation is so fast that it is unlikely that any actual calculation is involved. However, the fact that *future* dates are ascertainable and that they are, presumably, only rarely available for memorization makes this seem an unlikely general explanation.

An additional argument used to support the rote-memory hypothesis is based on the belief that *idiot savant* calendrical calculators lack insight into how they accomplish their feat. This is seen as consistent with the view that they do not have the capacity for abstract thought, which, in turn, is used as circumstantial evidence that rote memory, rather than thought, is the basis of their talent. However, O'Connor and Hermelin (1984) have shown that calendrical calculators have some knowledge of the rules and regularities of the calendar (see also Hermelin and O'Connor 1986). Indeed, it was one of their subjects who first drew their attention to the rule that, within any century, years which are 28 years apart will have identical calendar structures (O'Connor, personal communication).

(3) There have been anecdotal suggestions that calendrical calculation may be based on visual images, or representations, of actual calendars. Sacks (1985) reported the impression that the twins Charles and George were reading numbers from some kind of stored image. Howe and Smith (1988) noted that the calendrical calculator in their study actually spent a fair amount of time drawing calendar months. Inspection of the drawings suggested that they were copies of an actual calendar in the young man's home. Indeed, he would sometimes volunteer the information that particular days – for example, Thursdays – were black; and when pressed as to why, he would reply 'They're black on the kitchen calendar' (Howe and Smith 1988, p. 9).

It is likely that the calendrical calculators do not all use the same method for making their calculations. Some may rely on memory, others on knowledge of the calendar, and others on visual imagery. It is doubtful, consequently, whether this group is homogeneous with respect to their cognitive architecture. But what are the possibilities suggested by the theory of the minimal cognitive architecture?

Given the characterization of modules in chapter 4 (dedicated, complex computational systems designed to acquire specific knowledge that has survival value, from information that is evolutionarily invariant), it is unlikely that calendrical calculation has a modular basis.[4] On the other hand, Mark II modules are not complex computational devices in the sense of providing classes of mental representations using specific, dedicated computational algorithms; yet they are no less necessary for the processing of information. For example, it may be that, provided the necessary information is encoded in the first place, retrieval from long-term memory may be unrelated to individual differences in intelligence (N. R. Ellis 1963, Anderson 1988). As long as the efficiency of such a process does not co-vary with individual differences in intelligence, then it represents an independent, unconstrained aspect of the architecture which would have modular status within the theory. If calendrical calculating depends on one or more Mark II modules – for example, on a rote memory built up through extensive practice – then it may be better to think of this talent as a quirk of personality rather than cognition.

The other possibility is that calendrical calculators have a particularly powerful specific processor. This could be of either kind; but a reasonable hunch would be SP2, the processor claimed in chapter 5 to have spatial, rather than propositional, properties. It may be that these individuals have an SP2 that, when so motivated, can generate a powerful code of such an elegance and efficiency that it can be implemented even on a slow basic processing mechanism. For example, SP2 may be able to perform operations that simulate normal, propositionally based calculations by some sort of imagery process.

There are fragments of evidence which suggest that a specific processor basis for some calendrical calculators may be worth exploring. First, O'Connor and Hermelin (1984) note that, although the eight calendrical calculators in their study showed a phenomenal degree of speed and accuracy of response, there was a within-group correlation of 0.4 between speed of calculation and IQ. This may reflect the influence of the basic processing mechanism on even economic and elegant specific processor algorithms. Second, Hermelin and O'Connor (1986, p. 889) note that 'the ability to calculate dates in the more distant future and across centuries seems to depend on the subject's level of general cog-

---

[4] Unless a module, or some process of modularization, has undergone some kind of neuronal distortion that has changed its possible range of inputs (Johnson and Karmiloff-Smith, in press).

nitive ability' and that 'It thus seems that the kind of internal representation from which strategies arise is itself a function of general cognitive
development' (p. 893). Whatever the processes used by these calculators, they do not seem to be immune from the constraint of the basic
processing mechanism. This reinforces the argument against a modular
basis for calendrical calculating.

If it is the case that calendrical calculating, by contrast with language,
music, and art, is supported by a powerful specific processor, then this
should be testable. For example, if the specific processor involved is
SP2, then we might expect rates of mental rotation to be faster for *idiots
savants* than for IQ-matched retarded controls and even for normal
subjects (certainly after the effects of variations in the speed of the basic
processing mechanism have been controlled). In any event, within-
group calculating performance should be predictable from measures
such as IT.

We have considered the possibility that two kinds of mechanisms,
modules and specific processors, may be the basis of quite different
kinds of *idiot savant* talents. Spared modules may be the basis for linguistic and artistic talent, whereas specific processors may be at the root
of calendrical calculation. I would now like to explore the possibility
that the third mechanism in the minimal cognitive architecture, the
basic processing mechanism, can be selectively spared from brain damage, which would in turn allow for yet other kinds of *savant* talent.

The case of a prime-number calculator

M.A. is a young man of 21 who, at the age of 3, was diagnosed as autistic. Although a normal baby, he suffered convulsions at 10 months and
again, repeatedly, between 2 and 4 years of age. The convulsions are
assumed to have caused brain damage, although results from EEG
scans proved inconclusive. As a young child, M.A. did not talk or
attempt to engage in any kind of communication. Since then, he has
learned to copy numbers and letters, but only very poorly. He still cannot talk, and has learned only a few elementary Paget Gorman signs.[5]
His psychometric performance is very poor. He is unscorable on the
Peabody Picture Vocabulary test, and his non-verbal IQ on the
Columbia Mental Maturity Scale is 67. Yet there is one anomaly. His
Raven's Progressive Matrices performance is accurate (and fast), and
suggests that he has an IQ of 128! However, his Raven's score stands in
stark isolation relative to the rest of his cognitive performance, both

[5] Word-based signs, first developed for use by the deaf, that preserve the word order of
spoken English.

psychometrically and as witnessed by his everyday skills. Yet, from an early age it was noted that he did have some exceptional qualities. As a young child, he could do complex jigsaw puzzles, even with the pieces face down. He was good with money, time, maps, and calendars, and was known to be able to add, subtract, multiply, divide, and even factorize numbers. It may be significant that both his parents have degrees in mathematics.

Hermelin and O'Connor (1990) investigated the calculating ability of M.A. when he was 20. They compared his performance with that of a male psychologist who is also a research scientist with a degree in mathematics. Hermelin and O'Connor had M.A. and the control subject complete three calculating tasks: in the first, they had to factorize three-, four-, and five-digit numbers; in the second, they had to identify prime numbers from a list of primes and non-primes; and in the third, they had to generate all prime numbers between two stated numbers.[6] In the factorizing task M.A. made fewer errors than the control subject (20 per cent compared with 36 per cent), with the majority of errors for both subjects being made on numbers greater than 10,000. In the recognition task M.A. was more accurate at recognizing primes than the control subject, except for numbers greater than 10,000, where both approached chance responding. In the generating task the level of performance was similar for both subjects.

A couple of general points emerged from this study. First, although the accuracy of the two subjects was similar, M.A. was much faster at making his calculations. Remember that his accuracy is, in any case, remarkable, given that it is comparable to that of an academic with a mathematics degree. Second, the pattern of errors was similar in the two subjects. For example, both often thought that the same non-prime numbers were primes. This suggests that however M.A. was going about the task, it was similar to the way in which the control subject was approaching it. This is significant, because understanding precisely how M.A. recognizes whether a number is a prime is crucial to interpreting the nature of his cognitive architecture. *If* he is calculating – that is, performing arithmetic operations on the number to arrive at a solution – and, moreover, if he is using some of the strategies that the control subject said he used for rejecting numbers as primes (for example, if a number is divisible by 11, then the addition of the first and last digits of a three-digit number will add up to the middle digit), then his basic processing mechanism must be functioning normally. I say this

---

[6] The factors of a number are those integers that divide the number with no remainder. A prime number is a number that is only divisible by itself and one. The process of factorizing a number usually ends with prime number components.

simply because the arithmetic calculations involved in determining whether a number is prime entail, under anybody's descriptive scheme, *thinking*, and thinking is constrained by the basic processing mechanism. Such complex thought processes would not be possible unless the basic processing mechanism operates at least at normal speeds. This leads to the prediction that if M.A. is truly calculating, then his basic processing mechanism must be operating fairly normally, as might be indexed, for example, by a low IT. Of course, this logic depends on the assumption that such calculating involves thinking. It need not, however, at least if we exclude 'seeing' as a variety of thinking.

S. B. Smith (1983) quotes a great calculator (Wim Klein) as saying to him: 'It doesn't mean the same for you, does it, 3844? For you it's just a three, and an eight and a four and a four. But I say "Hi, 62 squared"' (Smith 1983, p. 5).[7] Many great calculators do not know how they accomplish their feats and feel that they can just 'see' relationships between numbers. So it was said of Alexander Aitken that 'He has told me that results "come up from the murk" and I have heard him say of a number, that it "feels prime", as indeed it was' (J. C. P. Miller, cited in Smith 1983, p. 266).

Both Horowitz and Sacks seem to adhere to a non-thinking explanation of the calculating abilities of the twins George and Charles. Horowitz considered that their calendrical calculation was too fast to allow time for the necessary calculations to be made. He therefore believed that they could instantly recognize or associate a date with a day based on an extensive rote memory of the calendar. Klein, although he did not deliberately set out to learn the multiplication table up to 100 by 100, claimed that he in fact learned it by repeated exposure, thus acquiring it in a different way from his deliberate learning of the logarithm table by heart. Such familiarity with numbers and number combinations would allow fast access to already stored knowledge; thus: 'It is logical if you know that 2,537 is 43 times 59 and you're doing a little show . . . and they ask you for 43 times 59, you *recognize* straightaway 2,537' (Smith 1983, p. 288, emphasis added). But however accurately the verbs 'see' or 'feel' or 'recognize' capture the phenomenology, a simple association between the visual (or, for that matter, auditory) form of a number and properties as complex as its factor structure is an unlikely alternative to true calculating.

Colborn was a nineteenth-century calculating prodigy who devised the 'Colborn Table', a table which allows the factors of any number to be calculated. Colborn had the ability to factorize any number without

[7] In the following discussion of calculating, I am indebted to Smith's (1983) seminal book.

being aware of how he was doing it; no doubt, he too 'felt' that a number was prime, or 'saw' the factors if it was not. It was only at the age of 9 that he woke in the middle of the night and told his father that he realized, at last, how he made the necessary calculations (Smith 1983). His father proceeded to write down the complex sequence of rules which are the basis of the Colborn Tables.

Klein's ability to recognize products, perhaps as a result of familiarity, understates his calculating ability. When asked to multiply 426 by 843, to which he answered 359,118 (correctly) in a few seconds, he gave this explanation of how he did it: I say, hey, 426 divided by 6 is 71, and 843 times 6 is 5,058; 5 times 71 is 355 [in this case thousand] and 58 times 71, by experience without calculating it, is 4118' (Smith 1983, p. 290).

For those of you who cannot follow the verbal description, this is what he did:

Problem: 426 × 843

=   (71 × 6) × 843
=   71 × (6 × 843)
=   71 × 5,058
=   (71 × 5) + (58 × 71)
=   355(000) + 4,118

=   359,118

His method of multiplication involves not only calculation, but considerable insight and problem solving too. In short, it requires *thought*.

It is clearly an important issue for our discussion of the cognitive architecture of *idiot savant* calculators whether the method they use to perform these feats of calculation is a form of *thinking* or, alternatively, involves some kind of recognition process based on an elaborate but nevertheless rote memory. With these possibilities in mind, Beate Hermelin, Neil O'Connor, and I conducted a follow-up experiment with the prime number calculator M.A. with a view to understanding the cognitive nature of his ability (Anderson *et al.*, in prep.).

M.A. was presented with a number less than 1,000 on a computer display, and his task was to decide, making any necessary calculations in his head, whether the number was a prime. Of the 108 numbers presented to him, half were primes. There were three classes of non-prime number in the experiment. The first were *easy* non-primes, numbers that are divisible by either 2 or 5. The second were *computable* non-primes, numbers that were divisible by 3, 7, and 11, but not by 2 or 5. Finally, there were *hard* non-primes, numbers that are divisible by 13, 17, and 19 but not by 2, 5, 3, 7, and 11. The numbers also varied in their size. A third of all the numbers were less then 200, a third were greater than 200 but less than 500, and the remaining third were greater than 500. Primes and non-primes were equally distributed in these categories. Precise measurement of his response time to each number was made. The response times of M.A. were compared with those of a control subject (a university technician with a degree in engineering and a knowledge of mathematics) and with those of computer models designed to simulate different ways of discovering whether a number is, or is not, prime.

A comparison between the performance of M.A. and the control subject revealed two main features: first, that M.A. was much more accurate (he made only two errors in the 108 trials, compared with 18 by the control); second, that M.A. made his decisions in about a quarter of the time taken by the control (see figure 8.1). Even though M.A. was so much faster than the control subject, there was a marked similarity in the pattern of their responses to different numbers. Most obviously, they both took longer to indicate that a number was a prime than to indicate that it was not a prime. They also both took longer on hard

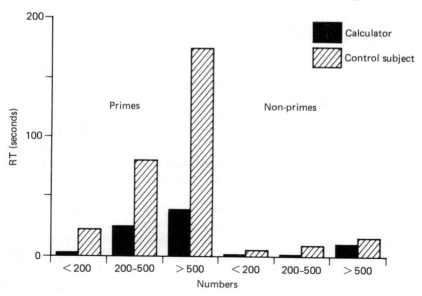

Figure 8.1 Comparison of M.A. (calculator) with the control subject for prime and non-prime numbers in the ranges <200, 200–500, and >500.

non-primes than computable non-primes, and both were very fast on easy non-primes. The fact that their pattern of response times was similar is significant, because the control subject admitted to a variety of calculating strategies. He would notice immediately whether a number was even or was divisible by five; if it was not, he would then try to divide it by three, then seven. Occasionally he would try the 'divisible by 11' strategy described earlier. He also said that he might make an attempt at division by higher primes, but towards the end, when he was fatigued, he tended to guess before he was sure. This last point is consistent with the fact that the majority of his errors were on numbers greater than 500, particularly those that were 'hard' non-primes.

To confirm the growing suspicion that M.A. recognized prime numbers by a process of calculation similar to that used by the control subject, his response times were compared with that of two computer simulations.

The simulations contrasted a true calculating process with a simple memory process, on the basis of the following rationale. If M.A. has a slow basic processing mechanism, then he may be recognizing prime numbers using a strategy that avoids the need to carry out complex computations of the sort that would be constrained by his slow processing speed. For example, such a strategy would be made possible by simply remembering lists of prime numbers. However, if we can establish that M.A. is calculating, rather than remembering, then we can predict that he cannot have the speed of processing typical of a mentally retarded person.

The memory model consisted of a stored list of prime numbers. When presented with a target number, the model searched sequentially through the list until the target matched a prime on the list or the target number was greater than the prime with which it was being compared. Simple assumptions about how long each comparison took were made, and response times to the 108 numbers were duly recorded. The basic features of such a model are that primes are faster to recognize than non-primes, and the larger the target number, the longer the response time.

The calculating model was based on an algorithm discovered by a Greek astronomer and mathematician, Eratosthenes, in the third century BC. If a target number is divided by all prime numbers less than its square root, and none divide without a remainder, then the target number is a prime number. The assumption of the Eratosthenes simulation was that a target number is divided successively by each prime, starting with 2 and ending with the last prime that is less than the square root of the target number. Again, assumptions were made about the speed with which each division was made, and the response times to each of the 108 numbers were recorded. The central features of this model are the opposite of those of the rote-memory model. Eratosthenes takes longer to identify primes than non-primes, and it is the lowest prime factor of a non-prime that determines the speed with which it will be recognized (rather than the absolute size of the non-prime target, as is the case for the memory model).

Because the absolute values of the response times of the simulations were arbitrary, the response times for each number category were converted to a z score for each simulation. The response times of M.A. were also converted to z scores. The transformations are presented in a bar chart, and represent the pattern of responses for each subject, controlling for between-subject differences in mean response time (see figure 8.2). The extent to which M.A.'s reaction times are more similar to the calculation, rather than the memory, simulation provides support for the hypothesis that M.A. is truly calculating and cannot, therefore, have the speed of processing normally associated with mental retardation.

Clearly the simple memory model does not fit M.A.'s data. The

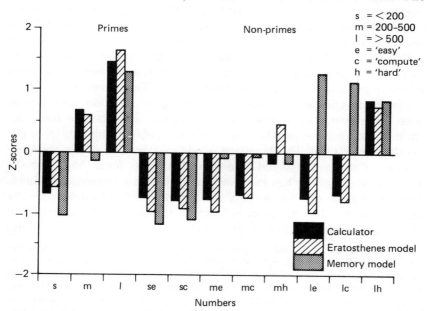

Figure 8.2 Comparison of M.A. (calculator) with simulations. Size of numbers: s = small; m = medium; l = large. Non-primes: e = divisible by 2 or 5 (easy); c = divisible by 3, 7, 11 (computable); h = divisible by 13, 17, 19 (hard).

memory model finds small numbers easy and large numbers difficult and primes easier than non-primes. However, even medium-sized primes are harder than all but large non-primes for M.A.

Both the Eratosthenes simulation and M.A. have the most trouble with large primes. Large computable (divisible by 3, 7, and 11) and large easy non-primes present little difficulty for Eratosthenes and M.A. In fact, M.A.'s responses shadow Eratosthenes over most categories. This suggests that M.A. is implementing a problem-solving algorithm, similar to that used by Eratosthenes, when deciding whether a number is or is not a prime. Certainly some form of calculation seems to be the basis of his decision. It is hard to imagine why a recognition system would take, for example, sixteen times as long to determine that 439 is a prime whereas 407 is a non-prime (407 is divisible by 11) or nine times as long to determine that 551 is a non-prime than it takes to determine that 561 is also a non-prime (561 is divisible by three, 551 by 19).

Given that M.A. seems to be performing calculations, and that calculating is a variety of thinking, it follows, since the calculations are difficult, that M.A. must have a basic processing mechanism that is functioning at, or above, normal levels of speed. It is cheering, then, that M.A.'s measured IT is approximately 35 milliseconds, significantly shorter than average.[8]

We can conclude that M.A. represents the opposite condition to D.H. D.H. has spared language modules but mentally subnormal levels of basic processing mechanism functioning. This prevents her from instantiating complex thoughts. So although she speaks fluently and with great syntactic, and apparent semantic, aplomb, she does not understand what she says because she is incapable of thinking what she so effortlessly says![9] Her language is not an extension of her thinking, because if it were, it would take a much simpler form. M.A., on the other hand, certainly has extensive damage to his linguistic modules, but has, in principle, the capacity to support complex thoughts. Yet the majority of his thought processes exhibit the poverty characteristic of the mentally retarded. It may be that the absence of linguistic modules

[8] His IT was measured using the space invader game described in chapter 7. An initial staircase algorithm estimated his IT at 30 milliseconds after 59 trials. MA was then given 60 trials, 12 each at exposure durations of 20, 40, 60, 80, and 100 milliseconds. He was at chance levels of accuracy at 20 milliseconds (58 per cent), but was 100 per cent accurate at 40 and 60 milliseconds (and made one error at both 80 and 100 milliseconds).

[9] Karmiloff-Smith (personal communication) believes that D.H. does understand what she says, and argues, instead, that D.H.'s comprehension failure occurs only in tests where she is required to disembody language from normal interactional speech – for example, when asked to match sentences with pictures.

from birth chronically handicaps thought, preventing a great deal of encyclopaedic knowledge from being acquired. However, it is clear that there are non-linguistic thought processes (see, for example, Weiskrantz (ed.) 1988); there are also cases reported of children reared without language who nevertheless develop knowledge of the world superior to that exhibited by M.A. But maybe M.A.'s problem is deeper than simply the absence of language; a re-examination of his psychometric profile leads to an intriguing possibility.

As already reported, M.A.'s Raven's Matrices score is very good. Certainly his level of performance requires abstract logic and deductive processes to a degree that would necessitate not only an efficient basic processing mechanism but also the ability to engage in symbol processing. Hermelin and O'Connor (1990) were provoked by the anomalous Raven's score into re-examining his Columbia Mental Maturity Scale performance. The latter requires the subject to pick the 'odd one out' of three, four, or five items. When the items were divided into those that were semantically based (the 'odd one out' distinction depending on semantic differences between the objects depicted; for example, a ladder being distinguished from a collection of objects that could be sat on) and those that were purely abstract or syntactic ('odd one out' of a number of shapes), it was found that he did quite well on the latter, but very badly on the former. It may be, then, that it is only in the situation in which a thought process requires reference to semantic properties of the real world that M.A. behaves as if he is mentally retarded. When the thought processes involve the abstract manipulation of symbols that make no particular reference to semantic properties of the real world, then M.A. behaves like a syntactic engine, and one with *turbo power* at that. It is as if he has no problem manipulating symbols in some kind of language of thought (Fodor 1975) but is unable to make anything but the simplest reference from his thought to meaning in his world. His ability to grasp *mathematical* concepts, but little else, may reflect the fact that mathematics itself is a reflection of the syntactic properties of our own thought processes. Unfortunately, having a fast basic processing mechanism may allow M.A. to make mathematical computations, but it is not enough to save him from mental retardation if other vital mechanisms are damaged. As I have said many times in this book, there is more to intelligence than processing speed.

Patterns of Abilities

I have concentrated on the attempt to explain *idiot savant* talents for two reasons. The first is that *idiots savants* are often claimed to refute one of the central propositions of the architecture – namely, that there are general properties of intelligence – and to support, instead, the contrary proposition that intelligence is a collection of independent abilities. Whether or not you have been convinced by my particular account of *idiots savants*, I hope to have established that there is no fundamental conflict between the proposition of general intelligence, as viewed from the minimal cognitive architecture, and the existence of *idiots savants*.

The second reason is that *savant* talents are inexplicable in terms of any other general theory of intelligence. Previous explanations of *savant* talents rely, typically, on the reporting of biographies to demonstrate likely sources of the talent in atypical childhood experiences (for example, intense exposure to music), or appeal to features of the individuals that may be atypical (for example, that they manifest obsessive characteristics or that they have some kind of rare, and unspecified, brain damage). In either case such explanations are almost entirely *ad hoc,* and make no contact with any general theories of intellectual performance. I want to claim that the minimal cognitive architecture at least offers a *principled* account of *savant* talents, and that, by doing so, it removes the phenomena from the category of fascinating oddities into the test bed of general theories of cognition.

The analysis of *savant* talents has been a limits-testing exercise for the theory. However, the theory can be applied to less extreme examples of cognitive specificity. An important class of these are cases of specific cognitive deficits. I have referred to such deficits in various places throughout the book, and it would take us too far from our main objectives to provide an analysis of the many putative specific deficits of interest to developmental psychologists. Further, I do not want to make the mistake of rushing off prematurely to develop particular models of specific deficits that attempt to fit the data generated by prior, highly developed models. This is because, as detailed models of specific abilities have developed, they have done so in a 'lop-sided' fashion, concentrating on circumscribed phenomena without the backdrop of the general picture that this book tries to provide. Almost certainly, then, much of the data will turn out to be model-specific and of no general significance, at least on the view of the intellect proposed here. However, there are clearly a number of specific deficits whose existence and significance do not depend on a particular model of cognitive development. A brief cameo of a couple of these should give some indication of the potential of the theory for explaining the relationship between specific deficits and intelligence.

Other Learning Disabilities

Dyslexia

There can be no doubt that some children have great difficulty in learning to read, write, and spell. 'Dyslexic' is a label commonly attached to such children, particularly if their reading problems are greater than we would expect, given their general level of intelligence or experience of reading. But such a statement hides a veritable morass of issues. Most obviously, the implication of *unexpected* reading failure as definitional is not to everyone's taste. I do not want to become ensnared in the many disputes in the literature, because such issues have been discussed at length by others better qualified than myself. Rather, and in the same way that we examined *idiots savants*, I would like to put forward a kind of *in principle* account of the nature of dyslexia, purely from the broader perspective of the theory. Dyslexia is a prime candidate for this kind of analysis, simply because, on everyone's story, intelligence is a particularly important moderating variable of reading performance. It is tempting, therefore, to use the theory to tease out what may be a number of intertwined relationships.

It is clear from the minimal cognitive architecture that there could be at least three kinds of reading failure.

(1) *Reading failure may be caused by a slow basic processing mechanism.* This kind of general reading failure will be predictable from levels of general intelligence. In some sense, of course, the term is a misnomer, for this kind of reading failure simply reflects the low general ability of the child to learn *any* new cognitive skills, as well as the impoverishment of the knowledge base upon which reading comprehension depends. Children with general reading failure would, of course, have a lower reading age than would be predicted from their chronological age, but this discrepancy would be predicted by their IQs. Note, however, that the relationship with IQ should not be regarded as being merely statistical (see van der Wissel and Zegers 1985, Frith 1985, Yule 1985, for a debate on this issue); rather, it is one of shared causality. Children with higher IQs *should* have higher-than-average reading ages, because both depend, in large part, on the speed of the basic processing mechanism. If a child with a high IQ (more properly, with a fast basic processing mechanism and powerful specific processors) is below average in reading, then *something else is wrong* (see figure 8.3). This should make clear that the concept of unexpected reading failure, as used in this context, is mechanistic and not merely statistical.

(2) *Reading failure may be caused by a poor specific processor.* I will call this condition 'specific reading failure' to distinguish it from the general kind described above.[10] We have seen in chapter 5 that we could think of the two specific processors as knowledge-acquisition mechanisms utilizing two different modes of processing, one more suited to verbal (sequential/propositional), the other to spatial (simultaneous/analogue) material. If either of the specific processors is particularly poor, this may result in poor reading skills in so far as reading depends on that mode of processing. It is unavoidable, of course, that specific reading failure will also be related to IQ, since the manifest power of each specific processor is, in part, a function of the speed of the basic processing mechanism, and all three mechanisms contribute to IQ differences. However, specific reading failure will be differentiated from general reading failure in the following ways:

(a)   Most obviously, general reading failure will be highly correlated with indices of processing speed, such as IT, whereas specific reading failure will be better predicted by measures that reflect specific processor functioning. Those might be either 'verbal' processing measures, such as, for example, the speed at which letters or digits are searched in short-term memory (S. Sternberg 1966) or 'spatial' processing measures, such as, for example, mental rotation (Shepard and Metzler 1971). Perhaps the process differences indicated by these measures will distinguish between different 'types' of dyslexic, although this begs another thorny question! At any rate, there does seem to be strong evidence that more specific cognitive processes, analogous to those which may characterize operations of the specific processors, are implicated in skilled reading. For example, Baddeley (1986) has argued that skilled reading is influenced by the capacity of an articulatory loop, which maintains phonological information in working memory and is particularly concerned with processing information about order.

(b)   Because, *ex hypothesi*, specific reading failure is caused by a poor specific processor, there must be other cognitive consequences too. Let us make the simplistic assumption that skilled reading depends more on the sequential/propositional processes typical of SP1 algorithms. On average, a poor SP1 will lead to particular problems with reading.[11] However, it is the nature of specific processors that each can,

[10] This is not to imply that the problem will not manifest itself in other cognitive deficits, as should become obvious on reading this section.

[11] Note that how specific to reading these problems are will be a consequence of how much reading, as opposed to other cognitive abilities, depends especially on SP1 algorithms. Clearly, there are likely to be other cognitive abilities that depend as much, if not more, on SP1 processes.

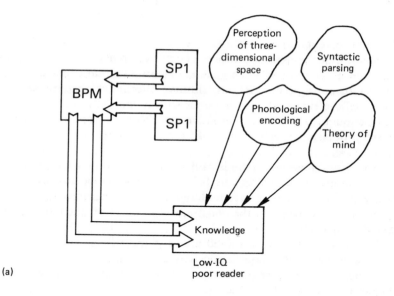

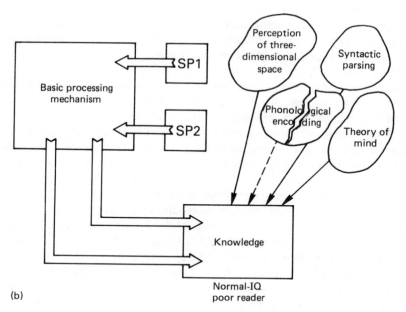

Figure 8.3 *A slow basic processing mechanism will lead to a poor knowledge base, and poor reading ability can be expected, as in (a). However, if a fast basic processing mechanism and rich knowledge base accompany poor reading ability, then something else is wrong: perhaps a damaged module, as in (b). BPM = basic processing mechanism.*

to some extent, compensate for deficiencies in the other. What this means is that the extent to which a cognitive ability *depends* on SP1 is a function of whether SP2 can generate algorithms that can substitute for the more appropriate SP1 algorithms. Alternative reading strategies, then, will be expected in individuals with a poor SP1 but an SP2 powerful enough to generate alternative algorithms for reading and with a basic processing mechanism with sufficient speed to run them (see figure 8.4).

So general reading failure is caused by a slow basic processing mechanism, whereas specific reading failure, although associated more with a deficient SP1, will still be moderated by processing speed (roughly speaking IQ) because of the constraint it imposes on the use of alternative SP2 algorithms.

The kind of pattern outlined above is most likely to manifest itself psychometrically in a correlation between reading problems and a discrepancy between verbal and performance IQ, the latter being much higher than we would predict from the former in children with specific reading problems.

Although the distinction between general and specific reading failure is an old one, my guess is that specific reading failure is often confused with the third kind of possible reading failure.

(3)   *Reading failure may be caused by a defective module.* This is the category of disorder that, from the perspective of the theory, best approximates 'pure' dyslexia. Having an important module that is implicated in normal reading damaged or missing, such as a mechanism responsible for phonological encoding or segmentation (Stanovich 1986, Frith 1986, Liberman 1983), would have profound effects on reading, at least for alphabetic scripts.

Developmental studies of dyslexia have been influenced, since the early 1980s, by neuropsychological studies of acquired dyslexia (A. W. Ellis 1985) – that is, cases in which brain damage has caused reading problems in previously literate adults. Using this framework, two kinds of dyslexia have been identified. The phonological dyslexic has lost the ability to decode graphemic symbols into their associated sounds. This kind of dyslexia is characterized by an inability to read invented nonwords (for example, 'greep') and the preserved ability to read irregular words (words whose spelling violate the normal grapheme–phoneme conversion rules, such as 'gauge') with which the person was familiar before the injury. The *surface* dyslexic, on the other hand, can 'sound out' words, so can read invented non-words, but not irregular words. For example, he or she might find it impossible to read 'yacht' or might

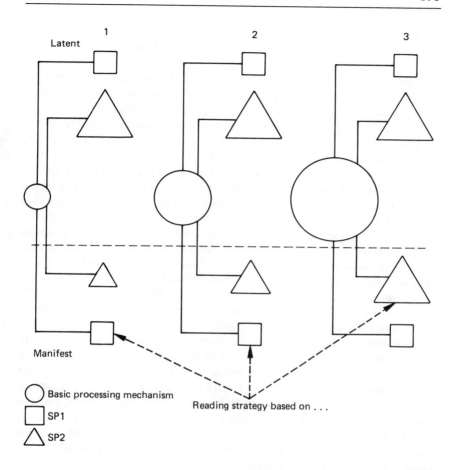

Figure 8.4 *Reading strategy as a function of the speed of the basic processing mechanism and the differential latent ability of SP1 and SP2. Normal reading is based on SP1 unless its manifest power is substandard and the manifest power of SP2 is superior, as in 3 above. Example 1 would be classified as general reading failure (slow speed of processing) but example 2 may be a case of a specific reading deficit due to poor manifest power of SP1 but normal speed of processing.*

mispronounce 'pint' as a rhyme for 'mint'. The analogy between acquired and developmental dyslexia has been embraced enthusiastically (Marshall 1984, A. W. Ellis 1985), and developmental equivalents of these two types have been discovered (Temple and Marshall 1983). While some have questioned the status of the evidence (Bryant and Impey 1986) and others have questioned the plausibility of a developmental disorder mirroring an acquired deficit in an adult (Frith 1986), there is little doubt that this approach is viewed as offering the key to discovering the fundamental deficits that cause dyslexia.

The *lack* of an ability or a competence, as implied by the application of the acquired dyslexia framework to developmental dyslexia, lends itself more to a modular explanation than to the continuum of disability intrinsic to the specific processor explanation of specific reading failure outlined above. However, given that literacy is a relatively recent human accomplishment, it would be facile to claim, for example, that the ability to map phonological codes onto graphemic codes is afforded by a biologically given module in the same sense as, say, ecological visual perception is. Of course, the story of modules may be expanded to include modules Mark III, those with an ontogenetic rather than a phylogenetic basis (see Karmiloff-Smith 1986, in press, and Johnson and Karmiloff-Smith, in press, for a discussion of the ontogenetic process of modularization). However, this is not the time to attempt a last-minute expansion of the theory! Instead, what we could look for is the source of dyslexia in a module that evolved to fulfil a purpose quite different from the reading of alphabetic scripts. Phonological dyslexia offers just such a possibility (see figure 8.3 (b)).

Phonological dyslexia may be, in essence, a speech problem, and the perception of speech seems to fit the criteria, outlined in chapter 4, for functions likely to be based on modules (Mark I). It may be that the phonological representations of speech in such children do not afford the mapping with graphemic symbols that is a necessary part of a major route to reading. Alternatively, their problem may lie in other mechanisms whose job it is to segment the speech representation into its phonemic components (possibly more like modules Mark II, the fetch-and-carry operators of information processing). The consequence is likely to be the same in either case, and the fact that 'dyslexic' children have difficulty segmenting speech is consistent with these speculations.

The modularity thesis for reading, and indeed for speech (Massaro 1989), is a matter of current contention, but from the perspective of the theory of the minimal cognitive architecture, if dyslexia can occur at all levels of intelligence, which does seem to be the case, then the deficit must be modularly based, *by definition*. The availability of alternative reading strategies (that is, problem-solving algorithms that avoid the need to use representations normally afforded by the damaged module) will probably allow average and maybe even good readers to exhibit this kind of dyslexia.

It should be possible, then, in the case of modular damage, for children with high IQs to be dyslexic. Further, and in an analogous fashion to the linguistic *savant* D.H. described above, children of low IQs have been found who, nevertheless, are extremely good readers. Just as D.H. seems not to understand what she says, so these 'hyperlexic' children fail to understand what they read (Snowling and Frith 1986).

So much for the attempt to disentangle dyslexia. Clearly, for the hypotheses to be firmed up, we need to make connections between, on the one hand, the cognitive processes involved in reading and particularly reading development (see Frith 1986 for a discussion) and, on the other hand, the mechanisms of the minimal cognitive architecture. But at the very least I hope that the theory provides a new backcloth against which any specific deficit in reading will be more clearly identified. Bryant and Impey (1986), when evaluating claims that two young adults were phonological and surface dyslexics respectively, asked the question 'To whom should the two dyslexics be compared?' and answered, 'to normal children of the same reading age' (p. 124). I hope that my speculations on the nature of dyslexia have shown just how much more complex the concept of a normal reader is than simply someone whose reading age matches his or her chronological age (see also Stanovich *et al.* 1988).

Autism

Another cognitive deficit that may be illuminated by the theory is autism. Frith (1989) presents a convincing argument that many cases of autism are best characterized as involving a cognitive deficit. Although there is an intriguing side to Frith's thesis, in that autistics are said to lack 'central coherence' of thought, I want to focus on her view that many autistics lack a particular mechanism that allows people to understand that other people have mental states, like beliefs, desires, wishes, and intentions (Baron-Cohen *et al.* 1985, Leslie 1987, Leslie and Frith 1990). This condition has been summarized as a deficient 'theory of mind'.

Again, I do not want to engage the many complex issues in an area well plied by others. Instead, I merely want to emphasize that the ability to understand that others have mental states seems, in the normal population, to be unrelated to intelligence; even mentally retarded children can succeed at tasks that require them to impute mental states to others, whereas autistic children of higher MAs and IQs often fail them. This would seem to argue that such a competence is modularly based, a view held by Leslie (Leslie 1987, Leslie and Frith 1990, Leslie 1991) for quite different reasons. Certainly, the absence of such a competence would have profound effects on a child's cognitive performance and may explain his or her lack of social interaction, attributed by others to a failure in understanding emotion (Hobson 1990, 1991).

What can the theory of the minimal cognitive architecture add to this story? Clearly, if the autistic deficit is truly modular, then, in terms of route 1 knowledge acquisition (that supported by the basic processing mechanism and the specific processors), many autistic people could, in fact, be considered intelligent. Low IQ, which is the norm for autistic people, would be attributable to the importance of the modular system underlying 'theory of mind' for the development of knowledge in general. For example, an inability to understand that others have mental states is likely to have profound and deleterious consequences for an autistic child's ability to learn from others. Further, if the 'theory of mind' deficit underlying autism has a wider basis, such as an inability to form metarepresentations (Karmiloff-Smith, in press), the implications for the organization of knowledge are likely to be catastrophic. Given the hypothesis of a modular deficit underlying autism, it is not without significance that in pilot work conducted in collaboration with Uta Frith, we have found many autistic people to have ITs very much lower than those of other mentally retarded people of comparable MA and IQ. Indeed, their measured ITs are of the level normally found in subjects of well-above-average IQ. This is consistent with the idea that the mental retardation associated with autism may be due, unusually, to a damaged module, rather than to slow processing speed. Indeed, M.A., the prime-number calculator described above, is another example of an autistic person with apparently fast processing speed, as is N.P., a musical *idiot savant* studied by Sloboda *et al.* (1985).[12]

Since speculation is now beginning to heavily outweigh the available data, it is time to end the discussions of how the architecture may be used to explain specific deficits and abilities and move on to formulating, more precisely, the regularities in the patterns of abilities that we see at different levels of intelligence and at different ages in development.

A General Formulation of Patterns of Abilities

In this chapter we have seen the model applied to particular examples of specific cognitive deficits and abilities. The general context of this

---

[12] I tested N.P. using my IT task in the early summer of 1990. His inspection time was only a little longer than that of M.A., the prime-number calculator, being approximately 55 milliseconds.

book makes clear, however, that this may be putting the cart before the horse. As I indicated at the beginning of the last section, I believe that there are problems in rushing too soon into specific models of particular deficits. We may spend a great deal of time refining a model to fit a particular case which might turn out not to be all that interesting when seen in the light of the general properties of the theory. To finish this chapter, I want to begin the groundwork for a general formulation of patterns of abilities. Such a pattern will provide a more sophisticated background against which models of specific abilities can be assessed. That is, it is more sophisticated than simply acknowledging that specifically handicapped individuals must be matched for 'intelligence' with normal individuals for any differences between the groups to be interesting. Matching for intelligence could be done more precisely by taking into account the three interrelated facets of the theory:

1   IQ differences are a function of route 1 knowledge-acquisition mechanisms; specifically, the speed of the basic processing mechanism and the power of the specific processors. Simply matching for IQ may obscure differences in the balance of strengths and weaknesses among these mechanisms that can lead to the same IQ score.

2   Similarly, matching for MA obscures the fact that MA reflects knowledge elaboration, which depends, in turn, on the efficiency of route 1 knowledge-acquisition mechanisms and how much experience a child has had (estimated by CA). Equal MAs may be achieved by a variety of combinations of these two factors.

3   In addition, MA is a cocktail of route 1 and route 2 (which depends on modules) knowledge acquisition. Although MA roughly equates with developmental level (see chapter 6), it confounds the relative contribution of these two routes to development. To pursue the cocktail analogy, matching for MA probably ensures that we are not comparing spirits with beers, but it may gloss over the distinction between a tequila sunrise and a malt whisky. No amount of experience and no level of IQ can compensate for some kinds of cognitive competences that have yet to, or may never, develop.

To make use of the theory for understanding patterns of abilities we first must begin to formulate, more precisely, the relationship between the mechanisms in the architecture. Indeed, I hope that the general formulation will provide ways of defining what constitutes a specific deficit.

Fitting the Parts Together

The central proposition of the theory is that intelligence is a property of thinking. In chapter 5 I proposed that when someone is thinking, he or she is running an algorithm generated by one of the specific processors. The faster the basic processing mechanism is, the more complex the algorithm that can be implemented. How complex a thought can be, therefore, is determined by two things: the latent power of the specific processor and the speed of the basic processing mechanism. The major hypothesis concerning the pattern of abilities that such an architecture should generate is that, since individual differences in intelligence are caused primarily by variations in the speed of the basic processing mechanism, at higher levels of intelligence, cognitive abilities should be more differentiated.

The differentiation hypothesis

Psychometrically, the differentiation hypothesis suggests that the general factor should decrease in importance (the amount of variation it accounts for in an intelligence test battery would be lower) as the intelligence of the group being tested increases. Concomitant with the decrease in importance of *g*, other more specific abilities should account for an increasing proportion of the variance. Just such an effect was claimed by Spearman (1927), who labelled it the 'law of diminishing returns' (see Deary and Pagliari, 1991). There are a number of lines of evidence that suggest, at least, that the structure of abilities changes with level of intelligence.

1   There is evidence that the general factor accounts for more of the variation among mentally retarded people than in people of average intelligence, a finding consistent with the differentiation hypothesis. As we saw in chapter 5, Spitz's (1982) construct of MA lag reflects the fact that it is on intelligence test subtests that are highly *g*-loaded that mentally retarded subjects do worse, compared with normal groups matched for MA. However, Spitz's hypothesis only partially maps onto the differentiation hypothesis. Since, for Spitz, what determines low IQ is primarily low *g*, then there will be an *inverse* relationship between the *g* loading of a test and intelligence in low-ability groups. Spitz (1988) confirmed this effect, and also argued that its corollary is that there should be a *direct* relationship between a subtest's *g* loading and performance in a group selected for higher-than-average intelligence. Since it is *g* that, hypothetically, determines intelligence, intelligent subjects

should score well on subtests that are good measures of *g*, but regress to the mean on those that are not highly *g*-loaded. This means that the proportion of variation attributable to *g* in high-ability groups should, as with the mentally retarded, be higher than average. It is on this latter point that Spitz's hypothesis deviates from that of the differentiation hypothesis proposed here.

2   The relationship between Verbal and Performance IQ changes with level of intelligence (Lawson and Inglis 1985). As Full-Scale IQ from the WAIS increases, the proportion of normal individuals with higher Verbal IQ than Performance IQ increases, and these two scales are most similar at average levels of ability (Matarazzo 1972).

3   It may be the case that, with the exception of *idiots savants*, already discussed, higher processing speed is, by and large, a prerequisite for exceptional talent. Terman in his classic five volume series *The Genetic Studies of Genius* regarded a high IQ as criterial for giftedness. While, clearly, this renders the link between 'intelligence' and talent one of definition, rather than observation, it should be seen in the historical context 'in which the "precocious" child was classed with the abnormals, depicted as a neurotic, and alleged, if he survived at all, to be headed for post-adolescent stupidity or insanity' (Terman and Oden 1947, p. 1). Further, the connection between talent and intelligence is well recognized even by those who have taken considerable care to distinguish between the two constructs (O'Connor and Hermelin 1981). For example, in a study of artistically and musically gifted children, when subjects were selected on the basis of their talent alone, their mean IQ was around 127. When divided into two groups on the basis of their IQ, even the 'lower IQ' groups were at least one standard deviation above the population mean (O'Connor and Hermelin 1983).

4   Recently, the differentiation hypothesis has found support, but from a different theoretical perspective: that of Detterman (1984, 1987). Detterman and Daniel (1989) divided the subject sample used to standardize the WAIS-R and WISC-R into those above and those below average IQ. The average intercorrelations among the subtests were much higher for the below-average-IQ group. This indicates a more pervasive influence of general intelligence at lower IQ (*ex hypothesi*, of processing speed) and greater specificity (*ex hypothesi*, increased influence of differences between the specific processors) at higher IQ. That these psychometric differences might be based on a different pattern of information-processing abilities was supported by another study conducted on a subsample of the subjects. Data from a battery of information-processing tasks (including choice reaction time and a form of IT) were more highly intercorrelated for a lower-IQ group than for a

higher-IQ group, and the data also correlated more highly with IQ within the lower group.

5   There has also been an 'age' version of the differentiation hypothesis. P.E. Vernon (1950, p. 29) cites the work of Garrett (1946), who claimed that the size of the g factor decreased with age, although, as Vernon points out, the older children in these studies were more selected for higher intelligence than the younger, so this could be a restriction of range effect rather than an age effect *per se*. Burt (1954) also considered that abilities became more differentiated, and therefore less g-dependent with age, reporting a decrease in the importance of the general factor (from 53 per cent to 48 per cent) and a rise in the importance of group factors (from 28 per cent to 47 per cent), between 9-year-olds and 14-year-olds. However, the age version of the differentiation hypothesis is particularly difficult to evaluate by psychometric test performance, because age comparisons on intelligence test performance are rarely made on the same battery of tests. This is a necessary consequence of the fact that different intelligence tests are designed for different age-groups, so as to maximize their discriminative power (for example, a test that *all* the older children succeed on is psychometrically useless). This makes it difficult to tell whether the structure of abilities has changed because of cognitive changes with age or whether it is an artifact of the fact that different abilities are being tested by different test batteries.[13]

There is, then, some support for the idea that the structure of abilities changes with level of intelligence. I will now try to make some of the patterns predicted by the minimal cognitive architecture a little more explicit by adopting a particular model of their relationship.

Up to now I have used diagrams to illustrate the general nature of the relationship between the different mechanisms of the minimal cognitive architecture. To conclude this chapter, I will summarize these relationships in terms of simple mathematical functions.

Simulating patterns of abilities

The most important relationship to capture is how the speed of the basic processing mechanism constrains the complexity of algorithms, generated by the specific processors, that can be implemented. We have seen this represented diagrammatically in figure 5.3. What we need now is a mathematical function that will allow the manifest ability of a

[13] This problem is being addressed by C. Pagliari in her doctoral thesis at Edinburgh University.

specific processor to rise as a function of the speed of the basic processing mechanism. We will assume this to be an asymptotic function in
conformity with the notions that (a) increases in processing speed make
the greatest difference at slow, rather than fast, speeds, and (b) there
will be an ideal state at the fastest processing speed, at which the manifest ability of a specific processor is unconstrained and is determined
only by its latent power. The most obvious asymptotic function to use
is the natural logarithm. Thus, we can make the manifest power of a
specific processor a logarithmic function of the speed of the basic processing mechanism. This relationship is illustrated in figure 8.5. In
addition, the manifest power of a specific processor increases as a function of its own latent power. The effect of increasing the latent power of
the specific processor can again be modelled as a logarithmic function,
as shown in figure 8.5. The major feature of the combination of the two
functions is that differences in the latent power of a specific processor
become more manifest at higher speeds of the basic processing mechanism.

It is now a simple matter to combine these functions to make 'cognitive
ability' a simple additive function of the manifest ability of the two

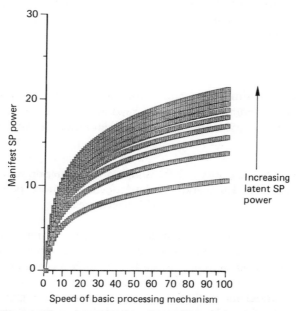

Figure 8.5 *The manifest power (y-axis) of a specific processor increases as a logarithmic function of both the speed of the basic processing mechanism (x-axis) and different
levels of latent power (shaded lines).*

specific processors. The central feature of the relationship between cognitive ability, the speed of the basic processing mechanism, and the latent power of the two specific processors is illustrated in figure 8.6. At lower cognitive ability, most of the variation is due to differences in speed of processing, whereas for higher cognitive ability, differences in the latent power of the specific processors become more important.

Figures 8.5 and 8.6 simply show us that the chosen mathematical functions embody the key features of the theory. What we can now do is to use the mathematical model to predict the likely pattern of abilities given (a) a putative 'model' of an ability and (b) estimated patterns of variation in the underlying mechanisms. Let me end this chapter by giving a simple (and simplistic) example of how we might apply this to our formulation of reading problems earlier in the chapter.

The relationship between IQ and reading problems

Figure 8.7 shows the relationship between IQ and reading performance for a population of simulated individuals. The assumptions underlying the model *vis-à-vis* intelligence are: (1) that the speed of processing and the latent power of the two specific processors are normally distributed and uncorrelated, (2) that the manifest ability of the specific processors is given by the function

$$SP_{manifest} = \ln (bpm_{speed}) \times \ln (SP_{latent}),$$

and (3) that IQ is simply the sum of the manifest ability of each specific processor. The assumptions *vis-à-vis* reading are that reading will be a function of the manifest ability of SP1, unless the following conditions hold:

(1) If the manifest ability of SP2 is higher than the population average *and* if it is at least one standard deviation more powerful than SP1, then an alternative SP2-based reading strategy will be adopted.
(2) Five per cent of the population (chosen at random) will be dyslexic. In these cases, reading performance will be based on the manifest power of SP1 reduced by 2 standard deviations.

This simple model shows that reading performance is highly correlated with IQ. This is not only a consequence of the fact that SP1 (on which normal reading is based in the model) and IQ are necessarily highly correlated. It is also due to the fact that there are a substantial number

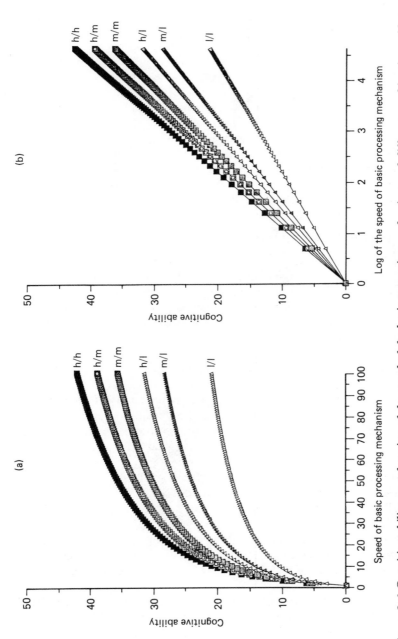

Figure 8.6 Cognitive ability as a function of the speed of the basic processing mechanism and different combinations of latent power the two specific processors: h = high; m = medium; l = low. The graphs show the increasing differentiation of specific abilities. In (b) the speed of the basic processing mechanism has been transformed to a logarithmic scale.

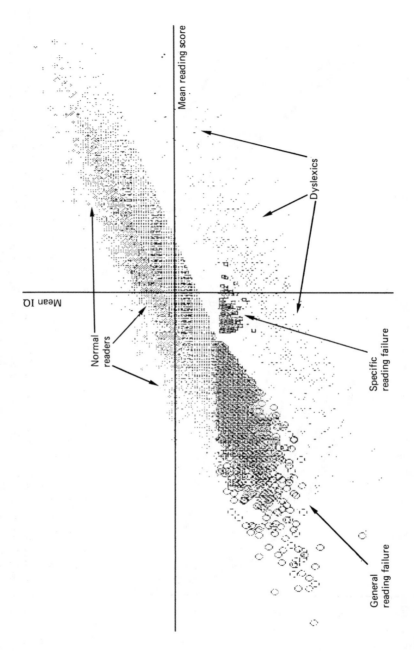

Figure 8.7 The relationship IQ and reading ability predicted by the model.

of alternative reading strategy-users (SP2-based reading) who, for the most part, must have higher-than-average IQ in order to be able to utilize the strategy. It is also true, paradoxically, that, as a group, strategy-users are better readers than normal (SP1-based) readers taken as a group! There are also a few individuals whose superior SP2 is just powerful enough to make them a little above average on reading, despite an obviously weak SP1, but whose manifest SP2 is not powerful enough to pull them up to average IQ. Most individuals with reading problems (general reading failure) are unremarkable, in that their reading performance is not out of line with their other cognitive abilities.

'Dyslexics' form a parallel 'band' beneath the non-dyslexic distribution. It is obvious that most of these readers, being of below-average IQ, would not be considered different from the general reading failure group. There is a reasonably large group who have higher-than-average IQ but who are still below average in reading. However, the discrepancy between their reading scores and the score 'expected' on the basis of their IQ is probably not substantial enough to allow them to be detected individually. It would only be when they were compared as a group with IQ-matched controls that significant differences in reading would emerge. It is instructive, however, for those who would ignore the relationship between intelligence and reading when investigating reading problems and concentrate instead solely on reading performance that a small number of 'dyslexics' are actually above average in reading.

This simple example illustrates the potential of the theory to make more accurate predictions regarding population distributions of cognitive abilities. The potential will only be fulfilled, of course, by constructing models of the cognitive abilities in question expressed in terms of the operation of the mechanisms of the minimal cognitive architecture. Supposing that this can be done, simulations like the one for reading, illustrated above, should be useful on at least two counts. First, it would allow us to refine the selection of control groups for experimental studies (for example, for most purposes we would want to select 'normal' good readers, not alternative-strategy good readers). Second, we should also be able to test predictions from different models (on the above model, the very best readers will be SP2 strategy-users, but SP1 strategy-users will constitute the bulk of good readers) using large population studies. Such studies would allow estimates of how 'important' (how much of the variance is accounted for) any particular variable is for the cognitive performance in question. Ultimately, however, the usefulness of this new approach to understanding the relationship between intelligence and specific cognitive abilities will depend on how good the underlying theory is.

9

Conclusion

In concluding, I want to do two things. First, I will summarize the theory and show how it accommodates the agenda I set at the beginning of the book. Second, I will try to pull a few strands together by showing how the theory relates to some wider issues in psychology.

The Nature of Intelligence and Cognitive Development

There are two routes to knowledge acquisition, which, roughly, cleave apart individual differences in intelligence and cognitive development.

> *Route 1 knowledge acquisition: Knowledge that is acquired by the implementation of an algorithm generated by a specific processor.*

The implementation of algorithms is constrained by the speed of the basic processing mechanism. When knowledge is acquired via this route, we can be said to be *thinking*. The basic processing mechanism represents a knowledge-free, biological constraint on thought, and is responsible for the phenomenon psychometricians know as general intelligence. What is more, the speed of the basic processing mechanism is unchanging with cognitive development. Variations in the mechanisms underlying route 1 knowledge acquisition are the basis of *individual differences* in intelligence.

> *Route 2 knowledge acquisition: Knowledge that is directly given by dedicated modules.*

Modules have been fashioned by evolution to provide information with evolutionary cash value that could not be provided by the mechanisms of thought. The computations of modules are not constrained by the speed of the basic processing mechanism and are therefore unrelated to individual differences in intelligence. The maturation of modules is the primary cause of cognitive development. Therefore, individual differences in intelligence and cognitive development are caused by different and unrelated cognitive mechanisms. However, individual differences in intelligence influence cognitive development through the operation of a secondary developmental process: *knowledge elaboration*.

Knowledge Elaboration and the Language of Thought

There are three influences on knowledge elaboration:

1   Knowledge is elaborated using route 1 knowledge-acquisition mechanisms. Individuals with more powerful route 1 mechanisms will have more elaborate knowledge structures. In this way, knowledge elaboration with development will be related to individual differences in intelligence.

2   Knowledge elaboration will also depend on the representational systems available to thought; such systems may be considered to be a *language of thought*. The maturation of a module may increase the power of this language of thought by providing alternative representational formats. In such a case, the maturation of a module would lead to a dramatic and universal increase in the cognitive competence of the child. The change in representational power would also provoke a new wave of knowledge elaboration.

3   Knowledge elaboration will be influenced by the range of experiences to which the developing child is exposed. *All other things being equal* (which we know they are not; but we know now what the relevant dimensions of difference are), knowledge elaboration will be a function of age.

In sum, sweeping changes in cognitive development, because they are based on modules, owe nothing to individual differences in intelligence, or experience, but everything to a biological programme of development which is roughly dependent on chronological age. However, the subsequent elaboration of this new competence in knowledge structures is once again constrained by the mechanisms underlying route 1 knowledge-acquisition; roughly put, a measure of individual differences in intelligence, such as IQ, will predict knowledge elaboration in development. Finally, because developmental change also depends on knowledge prerequisites – that is, the degree of elaboration of the different representational systems – developmental changes will be predicted best of all by the mental age of the child. This is because mental age is a cocktail of both modular change and knowledge elaboration.

The Agenda

In the course of this book we have, inevitably, gone way beyond our agenda. The attempt to accommodate the major phenomena of interest has produced a theory that addresses many more issues than those with which we started. This is as it should be. But in order to tidy up the story, we should return to the major issues with which we started. In chapter 1 we divided our agenda into two sections: the *regularities* and the *exceptions* to those regularities. I will deal first with three regularities and then with the two exceptions.

Regularities

1 *Cognitive abilities increase with development.* The developmental dimension of the theory proposes that there is a characteristic acquisition profile for diverse cognitive abilities because of the general increase in cognitive ability afforded by the maturation of new modules. There will be surges in underlying competence as new modules mature, but the process of knowledge elaboration will blur the steps in performance that would otherwise occur. Knowledge elaboration will be a function of both the efficiency of route 1 knowledge-acquisition mechanisms (that is, individual differences in intelligence) and experience. In turn, experience is likely to affect elaboration, with diminishing returns as a function of time. That is, most elaboration will take place soon after the maturation of a module and the effect of experience will reach an asymptote some time later; although precisely when this will be will vary with different abilities.

2 *Individual differences are remarkably stable in development.* The stability of individual differences is due to the unchanging speed of the basic processing mechanism.

3 *Cognitive abilities co-vary.* Cognitive abilities co-vary because all abilities that depend on thought are subject to the constraint of the speed of the basic processing mechanism.

Exceptions

4 *There are also specific cognitive abilities.* There is more to cognition than general intelligence. In particular, specific processors subserve

different modes of processing. Individual differences in the latent power of the specific processors will only become apparent, however, at higher speeds of the basic processing mechanism. This leads to the increased differentiation of abilities at higher levels of measured intelligence.

5 *There are cognitive mechanisms that are universal for human beings and which show no individual differences.* Modules are computationally complex processing devices that are available to all but the brain damaged. Because their computations are unconstrained by the speed of the basic processing mechanism, modular functions are unrelated to individual differences in intelligence. The availability of modules to all members of the species explains why the mentally retarded have the capability to perform what are, at one level of description, very complex computational procedures, yet find some 'easy' computational functions very difficult. Further, damaged modules may be the basis of many specific cognitive deficits. This is why some highly intelligent children may have enormous difficulty with cognitive problems – for example, spelling – well within the grasp of much younger and less intelligent children. The ability to compensate for such damage is likely to be extremely limited, but higher processing speed will be advantageous if the psychological functions normally subserved by modules are to be simulated by the specific processors.

Wider Issues

The substantive arguments that I wanted to put forward in this book have now been presented. I will end by setting out a few markers on topics that, because of space limitations, have only been alluded to so far. This may help to fill a few holes for the waverers, and may suggest new avenues of research for the persuaded.

Levels of Description

The theory proposed in this book was provoked by an attempt to synthesize in a single coherent theory the data on intellectual abilities from a wide gamut of psychological research. The resulting theory is avowedly cognitive. It stands in contrast to other theories of intelligence that we have come across in the course of the book. Eysenck (1988) is explicitly reductionist: intelligence is first and foremost about neurophysiology. Gardner (1983) argues that intelligence is, at one and the same time, biological, cognitive, and cultural. R. J. Sternberg (1984) argues that a cultural conception is fundamental to the construct. As for me, I agree, of course, that there are also biological and cultural levels of description of intelligence, but I see no reason to believe that these different levels of description are mutually constraining. For example, as I pointed out in chapter 4, it is not at all clear that the biological, cognitive, and cultural conceptions of intelligence, as outlined by Gardner, all map onto the same theoretical construct 'intelligence'. If they cannot be shown to do so, then there is no coherent sense in which the theory can be said to embrace all three levels of description.

It is a fiction, I believe, that there is an entity 'intelligence' that transcends any particular descriptive framework. So, while there is a physiological, a psychometric, a cognitive, and a cultural conception of intelligence, there is no particular reason or imperative to suppose that at the end of the day all these conceptions must map onto the same 'thing'. It is likely that the important theoretical distinctions at one level of description are unimportant at another. Would it really matter if it turns out that the differences in speed of the basic processing mechanism are caused by differences in neurotransmitter X rather than neurotransmitter Y? Not for the *cognitive* theory. Would it matter that SP1 *encodes* rather than *transforms* propositional representations? Well, it most certainly would for the cognitive theory, but I fail to see its significance for a cultural theory, which may seek to understand, for example, why some activities are viewed as intelligent in one culture but not in another.

The best research strategy is not, in my view, to search for a unified theory of all the manifestations of what is termed 'intelligence'. Rather, it is to produce well-articulated theories in significant, but clearly delimited, domains. If true, this means that it is wrong to claim that a purely cognitive theory of intelligence is incomplete (R. J. Sternberg 1984). Incomplete in what sense? The completeness of the theory is determined by how it accommodates the data and phenomena it has chosen to address, and its power is determined by how much better it does so than competing theories. I hope I have demonstrated that for the agenda set in this book (which I would argue contains the core agenda for a *psychological* conception of intelligence) only a cognitive theory is appropriate. I also hope you are persuaded that the one proposed here is the best on offer.

Culture and Cognitive Performance

In this book, I have not said enough about the influence of experience and culture on intelligence for everyone's taste. In part, this is because I have focused on the cognitive nature of intelligence, which, as I argued above, is a quite separate construct from its cultural manifestations. In part also, this is a reaction against the *Zeitgeist* that has prevailed in cognitive and educational psychology over the last twenty years which has tended to overemphasize the importance of experience relative to biology. Biology is obviously important in development, and cognitive theories of development should start to pay more than lip service to this fact. That said, the theory does make clear, although I have not emphasized this up to now, that experience is crucial for cognitive development. For example, it should be remembered that, modular development apart, the difference in intellectual performance between an infant and an adult is due solely to the elaboration of knowledge. The elaboration of knowledge, while constrained by processing speed, is dependent on experience.

That biological and experiential variables are both *necessary* contributors to intellectual development is, of course, a truism. But this is not to imply that these variables are inextricably confounded; their relative importance is a matter for empirical investigation. The techniques provided by modern behavioural genetics, coupled with a more precise theory of what constitutes a cognitive ability, should allow us to distribute variation in performance among these different causal bases.

Psychometrics

What does the theory suggest about the practice of intelligence testing?

Performance on an intelligence test is dependent on many factors, but, by and large, intelligence tests work because they reflect what individuals *know*. What individuals know is, in turn, a function of both the efficiency of the mechanisms responsible for acquiring knowledge and the range of available experiential inputs. The relative importance of these sources of variation for performance on any particular intelligence test is an empirical, rather than a theoretical, question. However, the theory outlined in this chapter would argue that an IQ score is, primarily, a *measure* (not a psychological attribute) of relatively stable differences between individuals in the efficiency of their knowledge-acquisition mechanisms.

An IQ score can be used as a measure, albeit an imperfect one, of the theoretical construct 'intelligence'. But there is a paradox here. IQ scores are *useful* in educational, occupational, and clinical settings, precisely because they are imperfect measures of the theoretical construct 'intelligence'. Even if we could measure, for example, processing speed *perfectly* and even though we hypothesize that variations in processing speed are the primary cause of individual differences in intelligence, we

would still use an IQ test in preference to a laboratory measure of processing speed for predicting everyday cognitive performance. I say this
for two reasons.

First, IQ tests are *also* sensitive to other cognitive and experiential
variables that influence how well a child will do at school, how well
someone will perform in their job, and how well a head-injured patient
will recover. Because of this, they give a general picture of the intellectual status of an individual relative to a relevant comparison group.
Thus they are *useful* not to the extent that they approximate a scientific
instrument but to the extent that they are good predictors of some performance criteria.

Second, cognitive measures (laboratory tasks developed for experimental investigations) will never replace psychometric tests of intelligence, because they lack the robust qualities (principally reliability)
needed for large-scale testing and comparison with population norms.
With a more specific theory of the expected pattern of abilities, IQ tests
should become even more useful than they have proved to be in the
past. However, if we have a theory about the mechanisms involved in
human intellectual functioning and we have some suspicion that in a
particular individual one of these mechanisms may be damaged, cognitive measures will provide a more refined, more specific diagnostic tool.
Information-processing tasks may assume the role of diagnostic tools
when testing for the relative strengths and weaknesses in cognitive abilities within the same individual. Just as GPs did not throw away their
stethoscopes and blood pressure meters when biochemical laboratories
opened for business, so an IQ test will always fulfil a necessary screening function in the diagnosis of intellectual disabilities.

Cognitive Science

The theory developed in this book addresses wider concerns in cognitive science. In particular, it argues that individual differences in intelligence are a property of thought, whereas development is a property of
mechanisms whose operation does not, in itself, constitute thinking.
This bears on the question which is at the very heart of the major controversy in cognitive science at the moment: Is the mind a symbol processor, or are symbols, rules, and representations merely convenient
descriptions – indeed, epiphenomena – of underlying neural networks?
Route 1 knowledge acquisition (thinking) clearly involves the processing of symbols; but whether modules are connectionist machines is
more debatable. However, at the very least, modules do not use the
same kind of symbol-processing machinery that thinking does.

There is, of course, a computational imperative in the theory that has not been realized. Although computational in spirit, the theory does not, as yet, embody a computational model. That is to say, the theory could not be used at the moment to explain precisely how someone solves even the simplest of problems. This may be the time to consider what would be the most fruitful line of theory development. In my view, developing full-blown computational theories for different cognitive abilities is still premature. The next stage must be to sketch broadly the computational characteristics of the different mechanisms. For example, is it the case that speed of processing constrains only *symbolic* (and not associative) processes? Experiments that contrast the relationships between intelligence and different kinds of cognitive processes and structures may zero in on the computational characteristics of the mechanisms faster than testing detailed models of particular cognitive abilities, such as reading or mental arithmetic.

The other side of the coin is that the framework provided by the theory of the minimal cognitive architecture allows us, for the first time, to use indices, such as IQ, to cleave apart different kinds of mechanism. The strong argument is that any psychological function that correlates with IQ is unlikely to be modular. Further, the relationship of different processes with measures of intelligence may help separate, empirically, those that are properly part of knowledge (*algorithms* for J. R. Anderson 1987, Pylyshyn 1980) from those that are part of its *implementation* (J. R. Anderson, 1987) or *functional architecture* (Pylyshyn 1980). It is not without considerable irony that IQ may turn out to be a valuable empirical tool in the cognitive scientist's methodological armoury.

Bibliography

Anderson, J. R. (1978) Arguments concerning representations for mental imagery. *Psychological Review*, 85, 249–77.

Anderson, J. R. (1987) Methodologies for studying human knowledge. *Behavioral and Brain Sciences*, 10, 467–505.

Anderson, M. (1986a) Inspection time and IQ in young children. *Personality and Individual Differences*, 7, 677–86.

Anderson, M. (1986b) Understanding the cognitive deficit in mental retardation. *Journal of Child Psychology and Psychiatry*, 27, 297–306.

Anderson, M. (1988) Inspection time, information processing and the development of intelligence. *British Journal of Developmental Psychology*, 6, 43–57.

Anderson, M. (1989a) The effect of attention on developmental differences in inspection time. *Personality and Individual Differences*, 10, 559–563.

Anderson, M. (1989b) Inspection time and the relationship between stimulus encoding and response selection factors in development. In D. Vickers and P. L. Smith (eds), *Human Information Processing: Measures, Mechanisms and Models*, North Holland: Elsevier Science.

Anderson, M. (1990) Intelligence and the development of intellect. *Spectrum, British Science News*, 222, 2–5.

Anderson, M., O'Connor, N. and Hermelin, B. (in prep.) Intelligence and cognition – A case study of a mentally retarded, savant prime number calculator.

Annett, M. and Manning, M. (1989) The disadvantages of dextrality for intelligence. *British Journal of Psychology*, 80, 213–26.

Baddeley, A. (1986) *Working Memory*. Oxford: Clarendon Press.

Baillargeon, R. (1986) Representing the existence and location of hidden objects: object permanence in 6 and 8-month old infants. *Cognition*, 23, 21–41.

Baron-Cohen, S., Leslie, A. M., and Frith, U. (1985) Does the autistic child have a theory of mind? *Cognition*, 21, 37–46.

Bayley, N. (1933) Mental growth during the first three years of life. *Genetic Psychology Monographs*, 14, 1–93.

Bishop, D. V. M. (1982) *T.R.O.G. Test for Reception of Grammar*. Abingdon, Oxon.: Thomas Leach Ltd.

Bishop, D. V. M. (1990) *Handedness and Developmental Disorder*. London: MacKeith Press.

Blinkhorn, S. F., and Hendrickson, D. E. (1982) Averaged evoked responses and psychometric intelligence. *Nature*, 295, 596–7.

Bornstein, M. H., and Sigman, M. D. (1986) Continuity in mental development. *Child Development*, 57, 251–74.

Brand, C. (1980) General intelligence and mental speed: their relationship and development. In M. Friedman, J. P. Das, and N. O'Connor, (eds), *Intelligence and Learning*, New York: Plenum.

Brand, C., and Deary, I. J. (1982) Intelligence and inspection time. In H. J. Eysenck (ed.), *A Model for Intelligence*, New York: Springer Verlag.

Brooks, J., and Weinraub, M. (1976) A history of infant intelligence testing. In M. Lewis (ed.), *Origins of Intelligence: Infancy and Early Childhood*, New York: Plenum.

Bryant, P. E., and Impey, L. (1986) The similarities between normal readers and developmental and acquired dyslexics. *Cognition*, 24, 121–37.

Bryant, P. E., and Trabasso, T. (1971) Transitive inferences and memory in young children. *Nature*, 232, 456–8.

Burt, C. (1954) The differentiation of intellectual ability. *British Journal of Educational Psychology*, 24, 76–90.

Cahan, S., and Cohen, N. (1989) Age versus schooling effects on intelligence development. *Child Development*, 60, 1239–49.

Carey, S. (1985) *Conceptual Change in Childhood*. Cambridge, Mass.: MIT Press.

Carpenter P. A., Just, M. A., and Shell, P. (1990) What one intelligence test measures: a theoretical account of the processing in the Raven Progressive Matrices test. *Psychological Review*, 97, 404–31.

Carroll, J. B., Kohlberg, L., and DeVries, R. (1984) Psychometric and Piagetian intelligences: towards resolution of controversy. *Intelligence*, 8, 67–91.

Case, R. (1985) *Intellectual Development: Birth to Adulthood*. London: Academic Press.

Ceci, S. J. and Liker, J. K. (1986) A day at the races: A study of IQ, expertise, and cognitive complexity. *Journal of Experimental Psychology: General*, 115, 255–66.

Chi, M. T. H. (1977) Age differences in memory span. *Journal of Experimental Child Psychology*, 23, 266–81.

Chi, M. T. H., and Ceci, S. J. (1987) Content knowledge: its role, representation and restructuring in memory development. *Advances in Child Development and Behavior*, 20, 91–142.

Chomsky, N. (1986) *Knowledge of Language: Its Nature, Origin and Use*. New York: Praeger.

Clark, H. H. and Chase, W. G. (1972) On the process of comparing sentences against pictures. *Cognitive Psychology*, 3, 472–517.

Cohen, L. B. (1981) Examination of habituation as a measure of aberrant infant development. In S. L. Friedman and M. Sigman (eds), *Pre-term Birth and Psychological Development*, New York: Academic Press.

Cohen, L. B., and Gelber, E. R. (1975) Infant visual memory. In L. B. Cohen and P. Salapatek (eds), *Infant Perception: From Sensation to Cognition*, vol. 1, New York: Academic Press.

Colombo, J. C., Mitchell, D. W., O'Brien, M., and Horowitz, F. D. (1987) The stability of visual habituation during the first year of life. *Child Development*, 58, 474–87.

Cooper, L. A., and Shepard, R. N. (1973) Chronometric studies of the rotation of mental images. In W. G. Chase (ed.), *Visual Information Processing*, New York: Academic Press.

Corballis, M. C. (1986) Fresh fields and postures new: a discussion paper. *Brain and Cognition*, 5, 240–52.

Courchesne, E., Hillyard, S. A., and Galambos, R. (1975) Stimulus novelty, task relevance and the visual evoked potential in man. *Electroencephalography and Clinical Neurophysiology*, 45, 468–82.

Cromer, R. F. (1974) The development of language and cognition: the cognition hypothesis. In B. Foss (ed.), *New Perspectives in Child Development*, Harmondsworth: Penguin.

Cromer, R. F. (1991) *Language and Thought in Normal and Handicapped Children*. Oxford: Blackwell.

Cromer, R. F. (in press) A case study of dissociations between language and cognition. In H. Tager-Flusberg (ed.), *Constraints on Language Acquisition: Studies of Atypical Children*, Hillsdale, N. J.: Lawrence Erlbaum Associates.

Das, J. P, Kirby, J. R., and Jarman, R. F. (1979) *Simultaneous and Successive Cognitive Processes*. New York: Academic Press.

Das Gupta, P., and Anderson, M. (in prep.) Intelligence and causal reasoning.

Deary, I. J., and Pagliari, C. (1991) The strength of g at different levels of ability: have Detterman and Daniel rediscovered Spearman's 'Law of Diminishing Returns'? *Intelligence*, 15, 247–50.

Dempster, F. N. (1985) Short-term memory development in childhood and adolescence. In C. J. Brainerd and M. Pressley (eds), *Basic Processes in Memory Development: Progress in Cognitive Development Research*, New York: Springer.

Dennis, M. (1985) Intelligence after early brain injury I: predicting IQ scores from medical variables. *Journal of Clinical and Experimental Neuropsychology*, 7, 526–54.

De Renzi, E. (1986) Current issues on prosopagnosia. In H. D. Ellis, M. A. Jeeves, F. Newcombe, and A. Young (eds), *Aspects of Face Processing*, Dordrecht: Nijhoff.

Detterman, D. K. (1984) Understanding cognitive components before postulating metacomponents, etc., part 2. Open Peer Commentary, *Behavioral and Brain Sciences*, 7, 289–90.

Detterman, D. K. (1987) Theoretical notions of intelligence and mental retardation. *American Journal of Mental Deficiency*, 92, 2–11.

Detterman, D. K. and Daniel, M. H. (1989) Correlations of mental tests with each other and with cognitive variables are highest for low IQ groups. *Intelligence*, 13, 349–59.

Donchin, E., and Coles, M. G. H. (1988) Is the P300 component a manifestation of context updating? *Behavioral and Brain Sciences*, 11, 357-74.

Egan, V. (1986) Intelligence and inspection time: do high-IQ subjects use cognitive strategies? *Personality and Individual Differences*, 7, 695-700.

Elliott, C. D. (1983) *British Ability Scales: Technical Handbook*. Windsor: NFER-Nelson.

Elliott, C. D. (1986) The factorial structure and specificity of the British Ability Scales. *British Journal of Psychology*, 77, 175-85.

Ellis, A. W. (1985) The cognitive neuropsychology of developmental (and acquired) dyslexia: a critical survey. *Cognitive Neuropsychology*, 2, 169-205.

Ellis, A. W. and Young, A. W. (1988) *Human Cognitive Neuropsychology*. Hove and London: Lawrence Erlbaum Associates.

Ellis, H. D., Jeeves, M. A., Newcombe, F., and Young, A. (eds) (1986) *Aspects of Face Processing*. Dordrecht: Nijhoff.

Ellis, N. R. (1963) The stimulus trace and behavioral inadequacy. In N. R. Ellis (ed.), *Handbook of Mental Deficiency*, New York: McGraw-Hill.

Ellis, N. R., Palmer, R. L., and Reeves, C. L. (1988) Developmental and intellectual differences in frequency processing. *Developmental Psychology*, 24, 38-45.

Ertl, J. (1966) Evoked potentials and intelligence. *Revue de l'Université d'Ottowa*, 30, 599-607.

Ertl, J., and Schafer, E. (1969) Brain response correlates of psychometric intelligence. *Nature*, 223, 421-2.

Eysenck, H. J. (1939) Review of *Primary Mental Abilities* by L. L. Thurstone. *British Journal of Educational Psychology*, 9, 270-75.

Eysenck H. J. (1967) Intelligence assessment: a theoretical and experimental approach. *British Journal of Educational Psychology*, 37, 81-98.

Eysenck, H. J. (1986) The theory of intelligence and the psychophysiology of cognition. In R. J. Sternberg (ed.), *Advances in the Psychology of Human Intelligence*, Hillsdale, N. J.: Lawrence Erlbaum Associates.

Eysenck, H. J. (1988) The concept of 'intelligence': useful or useless? *Intelligence*, 12, 1-16.

Fagan, J. F. (1971) Infants' recognition memory for a series of visual stimuli. *Journal of Experimental Child Psychology*, 11, 244-50.

Fagan, J. F. (1973) Infants' delayed recognition memory and forgetting. *Journal of Experimental Psychology*, 16, 424-50.

Fagan, J. F. (1976) Infants' recognition of invariant features of faces. *Child Development*, 47, 627-38.

Fagan, J. F. (1977) Infant recognition memory: studies in forgetting. *Child Development*, 48, 68-78.

Fagan, J. F. (1984) The intelligent infant: theoretical implications. *Intelligence*, 8, 1-9.

Fagan, J. F., and McGrath, S. K. (1981) Infant recognition memory and later intelligence. *Intelligence*, 5, 121-30.

Fagan, J. F. and Singer, L. T. (1983) Infant recognition memory as a measure of intelligence. In L. P. Lipsitt (ed.), *Advances in Infancy Research*, vol. 2, New York: Ablex.

Fantz, R. L. and Nevis, S. (1967) The predictive value of changes in visual preferences in early infancy. In J. Hellmuth (ed.), *The Exceptional Infant*, Seattle, Wash.: Special Child Publications.

Flynn, J. R. (1989) Rushton evolution and race: an essay on intelligence and virtue. *The Psychologist: Bulletin of the British Psychological Society*, 2, 363–6.

Flynn, J. R. (1990) Explanation, evaluation and a rejoinder to Rushton. *The Psychologist: Bulletin of the British Psychological Society*, 3, 199-200.

Fodor, J. A. (1975) *The Language of Thought*. New York: Thomas Y. Crowell.

Fodor, J. A. (1983) *The Modularity of Mind*. Cambridge, Mass.: MIT Press.

Frith, U. (1985) The usefulness of the concept of unexpected reading failure: comments on reading retardation revisited. *British Journal of Developmental Psychology*, 3, 15-17.

Frith, U. (1986) A developmental framework for developmental dyslexia. *Annals of Dyslexia*, 36, 69-81.

Frith, U. (1989) *Autism: Explaining the Enigma*. Oxford: Blackwell.

Gardner, H. (1983) *Frames of Mind: The Theory of Multiple Intelligences*. London: Heinemann.

Garrett, H. E. (1946) A developmental theory of intelligence. *American Psychologist*, 1, 372-8.

Geschwind, N., and Gallaburda, A. M. (1987) *Cerebral Lateralization: Biological Mechanisms, Associations and Pathology*. Cambridge, Mass.: MIT Press.

Gibson, J. J. (1979) *The Ecological Approach to Visual Perception*. Boston: Houghton Mifflin.

Gould, S. J. (1981) The *Mismeasure of Man*. Harmondsworth: Pelican.

Greenberg, D. J., Uzgiris, I. C., and Hunt, J. McV. (1970) Attentional preference and experience: III. Visual familiarity and looking time. *Journal of Genetic Psychology*, 117, 123–35.

Guilford, J. P. (1966) Intelligence: 1965 model. *American Psychologist*, 21, 20–6.

Gustafson, J. E. (1984) A unifying model for the structure of mental abilities. *Intelligence*, 8, 179-203.

Hadenius, A. M., Hagberg, B., Hyttnas-Bensch, K., and Sjogren, I. (1962) The natural prognosis of infantile hydroencephalis. *Acta Paediatrica*, 51, 117-18.

Haier, R., Robinson, D. L., and Braden, W. (1983) Electrical potentials of the cerebral cortex and psychometric intelligence. *Personality and Individual Differences*, 4, 591-9.

Halford, G. S. (1982) *The Development of Thought*. Hillsdale, N. J.: Lawrence Erlbaum Associates.

Halford, G. S. (1985) Children's utilisation of information: A basic factor in cognitive development. Draft paper, University of Queensland.

Halford, G. S. (1987) A Structure Mapping Analysis of Conceptual Complexity: Implications for Cognitive Development. Centre for Human Information Processing and Problem-Solving (CHIPPS): Technical Report 87/1.

Halford, G. S. and Leitch, E. (1988) Processing load constraints: a structure mapping approach. Paper presented at a symposium on Cognitive Development: Constraints, International Congress of Psychology. Sydney, Australia.

Halford, G. S., and Wilson, W. H. (1980) A category theory approach to cognitive development. *Cognitive Psychology*, 12, 356-411.

Halford, G. S., Maybery, M. T., and Bain, J. D. (1986) Capacity limitations in children's reasoning: a dual-task approach. *Child Development*, 57, 616-27.

Hasher, L. H., and Zacks, R. T. (1979) Automatic and effortful processes in memory. *Journal of Experimental Psychology: General*, 108, 356-88.

Hasher, L. H., and Zacks, R. T. (1984) Automatic processing of fundamental information: the case of frequency of occurrence. *American Psychologist*, 39, 1372-88.

Hendrickson, A. E. (1982) The biological basis of intelligence: Part II: Theory. In H. J. Eysenck (ed.), *Models of Intelligence*, New York: Springer Verlag.

Hendrickson, A. E. and Hendrickson, D. E. (1980) The biological basis for individual differences in intelligence. *Personality and Individual Differences*, 1, 3-33.

Hermelin, B., and O'Connor, N. (1986) Idiot savant calendrical calculators: rules and regularities. *Psychological Medicine*, 16, 885-93.

Hermelin, B., and O'Connor N. (1990) Factors and primes: a specific numerical ability. *Psychological Medicine*, 20, 163-9.

Hermelin, B., O'Connor, N., Lee, S., and Treffert, D. (1989) Intelligence and musical improvisation. *Psychological Medicine*, 19, 447-57.

Hick, W. G. (1952) On the rate of gain of information. *Quarterly Journal of Experimental Psychology*, 4, 11-26.

Hill, A. L. (1978) Savants: mentally retarded individuals with special skills. In N. R. Ellis (ed.), *International Review of Research in Mental Retardation*, vol. 9, New York: Academic Press.

Hindley, C. B. and Owen, C. F. (1978) The extent of individual changes in IQ for ages between 6 months and 17 years, in a British longitudinal sample. *Journal of Child Psychology and Psychiatry*, 19, 329-50.

Hissock, M., and Kinsbourne, M. (1987) Specialization of the cerebral hemispheres: implications for learning. *Journal of Learning Disabilities*, 20, 130-43.

Hobson, R. P. (1990) On acquiring knowledge about people and the capacity to pretend: response to Leslie (1987). *Psychological Review*, 97, 114-21.

Hobson, R. P. (1991) Against the 'Theory of Mind'. *British Journal of Developmental Psychology*, 9, 33-51.

Hofmann, M. J., Salapatek, P., and Kuskowski, M. (1981) Evidence for visual memory in the averaged and single evoked potentials of human infants. *Infant Behavior and Development*, 4, 401-21.

Horn, J. (1986) Intellectual ability concepts. In R. J. Sternberg (ed.), *Advances in the Psychology of Human Intelligence*, Hillsdale, N.J.: Lawrence Erlbaum Associates.

Horn, J., and Cattell, R. B. (1966) Refinement and test of the theory of fluid and crystallized intelligence. *Journal of Educational Psychology*, 57, 35-78.

Horowitz, W. A., Kestenbaum, C., Person, E., and Jarvik, L. (1965) Identical twin – 'idiots savants' - calendar calculators. *American Journal of Psychiatry*, 121, 1075–9.

Howe, M. J. A. (1990) *The Origins of Exceptional Abilities*. Oxford: Blackwell.

Howe, M. J. A., and Smith, J. (1988) Calendrical calculating in 'idiots savants': how do they do it? *British Journal of Psychology*, 79, 371–86.

Hulme, C., and Turnbull, J. (1983) Intelligence and inspection time in normal and mentally retarded subjects. *British Journal of Psychology*, 74, 365–70.

Humphreys, L. G., and Parsons, C. K. (1979) Piagetian tasks measure intelligence and intelligence tests assess cognitive development: a reanalysis. *Intelligence*, 3, 369–82.

Humphreys, G. W., and Riddoch, M. J. (1984) Routes to object constancy: implications from neurological impairments of object constancy. *Quarterly Journal of Experimental Psychology*, 36A, 385-415.

Hunt, E. (1978) Mechanics of verbal ability. *Psychological Review*, 85, 109–30.

Hunt, E. (1980) Intelligence as an information processing concept. *British Journal of Psychology*, 71, 449–74.

Hunt, E., and Lansman, M. (1982) Individual differences in attention. In R. J. Sternberg (ed.), *Advances in the Psychology of Human Intelligence*. Hillsdale, N. J.: Lawrence Erlbaum Associates.

Hunt, E., Lenneborg, C., and Lewis, J. (1975) What does it mean to be high verbal? *Cognitive Psychology*, 7, 194-227.

Hunt, J. M. (1974) Psychological assessment, developmental plasticity and heredity with implications for early education. In G. J. Williams and S. Gordon (eds), *Clinical Child Psychology*, New York: Behavioral Publications.

Hunter, M. A., Ames, E. W., Koopman, R. (1983) Effects of stimulus complexity and familiarization time on infant preferences for novel and familiar stimuli. *Developmental Psychology*, 19, 338–52.

Hyman, R. (1953) Stimulus information as a determinant of reaction time. *Journal of Experimental Psychology*, 45, 188–96.

Inman, W. C., and Secrest, B. T. (1981) Piaget's data and Spearman's theory – an empirical reconciliation and its implications for academic achievement. *Intelligence*, 5, 329–44.

Irwin, R. J. (1984) Inspection time and its relation to intelligence. *Intelligence*, 8, 47-65.

Jenkinson, J. C. (1983) Is speed of information processing related to fluid or to crystallized intelligence? *Intelligence*, 7, 91–106.

Jensen, A. R. (1969) How much can we boost IQ and scholastic achievement? *Harvard Educational Review*, 39, 1-123.

Jensen, A. R. (1980) Chronometric analysis of mental ability. *Journal of Social and Biological Structures*, 3, 181-224.

Jensen, A. R. (1982) Reaction time and psychometric *g*. In H. J. Eysenck, (ed.), *A Model for Intelligence*, Berlin: Springer Verlag.

Jensen, A. R. (1987a) The *g* beyond factor analysis. In R. R. Ronning, J. A. Glover, J. C. Conoley, and J.C. Witt (eds), *The Influence of Cognitive*

Psychology on Testing, Hillsdale, N. J.: Lawrence Erlbaum Associates.

Jensen, A. R. (1987b) Process differences and individual differences in some cognitive tasks. *Intelligence*, 11, 107–36.

Jensen, A. R., and Munro, E. (1979) Reaction time, movement time and intelligence. *Intelligence*, 3, 121–6.

Jensen, A. R. and Vernon, P. A. (1986) Jensen's reaction-time studies: a reply to Longstreth. *Intelligence*, 10, 153–79.

Johnson, M. H., and Karmiloff-Smith, A. (in press) Can neural selectionism be applied to cognitive development and its disorders? *New Ideas in Psychology*.

Joreskog, K. G., and Sorbom, D. (1981) LISREL V: Analysis of Linear Structural Relationships by Maximum Likelihood and Least Squares Methods. Research Report, Department of Statistics, University of Uppsala.

Just, M. A., and Carpenter, P. A. (1985) Cognitive coordinate systems: accounts of mental rotation and individual differences in spatial ability. *Psychological Review*, 92, 137–72.

Kail, R. (1986) Sources of age differences in speed of processing. *Child Development*, 57, 969–87.

Kail, R. (1987) Impact of extensive practice on speed of cognitive processes. Paper presented at the biennial meeting for research in child development, Baltimore, Md.

Kail, R. (1988) Reply to Stigler, Nusbaum and Chalip. *Child Development*, 59, 1154–7.

Karmiloff-Smith, A. (1986) From metaprocesses to conscious access: evidence from children's metalinguistic and repair data. *Cognition*, 23, 95-147.

Karmiloff-Smith, A. (1991) Beyond modularity: innate constraints and developmental change. In S. Carey and R. Gelman (eds), *Epigenesis of the Mind: Essays in Biology and Knowledge*, Hillsdale, N.J.: Lawrence Erlbaum Associates.

Karmiloff-Smith (in press) *Beyond Modularity: A Developmental Perspective on Cognitive Science*. Cambridge, Mass.: MIT/ Bradford Books.

Keating, D. P., and Bobbitt, B. L. (1978) Individual and developmental differences in cognitive processing components of mental ability. *Child Development*, 49, 155–67.

Kohlberg, L. (1969) Stage and sequence: the cognitive developmental approach to socialization. In E. E. Maccoby (ed.), *The Development of Sex Differences*, Stanford, Calif.: Stanford University Press.

Kosslyn, S. M. (1981) The medium and the message in mental imagery: a theory. *Psychological Review*, 88, 46-66.

Kosslyn, S. M. (1983) *Ghosts in the Mind's Machine: Creating and Using Images in the Brain*. New York: W. W. Norton.

Kosslyn, S. M. (1988) Aspects of a cognitive neuroscience of mental imagery. *Science*, 240, 1621–6.

Kranzler, J. H., and Jensen, A. R. (1989) Inspection time and intelligence: a meta-analysis. *Intelligence*, 13, 329–48.

Lally, M., and Nettelbeck, T. (1977) Intelligence, reaction time and inspection time. *American Journal of Mental Deficiency*, 82, 273–81.

Lally, M., and Nettelbeck, T. (1980) Intelligence, inspection time, and response strategy. *American Journal of Mental Deficiency*, 84, 553–60.

Lawson, J. S. and Inglis, J. (1985) Learning disabilities and intelligence test results: a model based on a principal components analysis of the WISC-R. *British Journal of Psychology*, 76, 35-48.

Lee, L. L. (1971) *Northwestern Syntax Screening Test*. Evanston: Ill.: Northwestern University Press.

Lenneberg, E. H., Nichols, I. A., and Rosenberger, E. F. (1964) Primitive stages of language development in mongolism. In D. McK. Rioch and E. A. Weinstein (eds), *Disorders of Communication*, Research Publications of the Association for Research in Nervous and Mental Disease, 42, Baltimore, Md.: Williams and Wilkins.

Leslie, A. M. (1986) Getting development off the ground: modularity and the infant's perception of causality. In P. van Geert (ed.), *Theory Building in Developmental Psychology*, North Holland: Elsevier.

Leslie, A. M. (1987) Pretense and representation: the origins of 'theory of mind'. *Psychological Review*, 94, 412–26.

Leslie, A. M. (1991) The theory of mind impairment in autism: evidence for a modular mechanism of development? In A. Whiten (ed.), *Natural Theories of Mind*, Oxford: Blackwell.

Leslie, A. M., and Frith, U. (1990) Prospects for a cognitive neuropsychology of autism: Hobson's choice. *Psychological Review*, 97, 122–31.

Levy, J., Trevarthen, C., and Sperry, R. W. (1972) Perception of bilateral chimeric figures following hemisphere disconnection. *Brain*, 95, 61-78.

Lewis, M., Jaskir, J., and Enright, M. K. (1986) The development of mental abilities in infancy. *Intelligence*, 10, 331–54.

Lewis M., and Brookes-Gunn, J. (1981) Visual attntion at three months as a predictor of cognitive functioning at two years of age. *Intelligence*, 5, 131–40.

Liberman, I. Y. (1983) A language-oriented view of reading and its disabilities. In H. Myklebust (ed.), *Progress in Learning Disabilities*, vol. 5, New York: Grune and Stratton.

Longstreth, L. E. (1984) Jensen's reaction time investigations of intelligence: a critique. *Intelligence*, 8, 139–60.

Longstreth, L. E. (1986) The real and the unreal: a reply to Jensen and Vernon. *Intelligence*, 10, 181–91.

Longstreth, L. E., Walsh, D. A., Alcorn, M. B., Szeszulski, P. A., and Manis, F. R. (1986) Backward masking, IQ, SAT and reaction time: interrelationships and theory. *Personality and Individual Differences*, 7, 643–51.

Mackenzie, B., and Bingham, E. (1985) IQ, inspection time and response strategies in a university population. *Australian Journal of Psychology*, 37, 257–68.

Mackintosh, N. J. (1986) The biology of intelligence? *British Journal of Psychology*, 77, 1-18.

Mandler, J. (1988) How to build a baby: on the development of an accessible representational system. *Cognitive Development*, 3, 113–36.

Marr, D. (1982) *Vision*. New York: Freeman.

Marshall, J. C. (1984) Multiple perspectives on modularity. *Cognition*, 17,

209–42.

Marshall, J. C., and Newcombe, F. (1966) Syntactic and semantic errors in paralexia. *Neuropsychologia*, 4, 169–76.

Martindale, C., Hines, D., Mitchell, L., and Covello, E. (1984) EEG alpha asymmetry and creativity. *Personality and Individual Differences*, 5, 77-86.

Massaro, D. W. (1989) Multiple book review of speech perception by ear and eye: a paradigm for psychological enquiry. *Behavioral and Brain Sciences*, 12, 741–94.

Matarazzo, J. D. (1972) *Wechsler's Measurement and Appraisal of Adult Intelligence*. Baltimore, Md.: Williams and Wilkins.

Mazur, J., and Hastie, R. (1978) Learning as accumulation: a reexamination of the learning curve. *Psychological Bulletin*, 85, 1256–74.

McCall, R. B. (1976) Toward an epigenetic conception of mental development. In M. Lewis (ed.), *Origins of Intelligence: Infancy and Early Childhood*, New York: Plenum.

McCall, R. B. (1979) The development of intellectual functioning in infancy and the prediction of later IQ. In J. D. Osofsky (ed.), *Handbook of Infant Development*, New York. Wiley.

McCall, R. B. (1981) Early predictors of later IQ: the search continues. *Intelligence*, 5, 141–7.

Mehler, J., Morton, J., and Jusczyk, P. W. (1984) On reducing language to biology. *Cognitive Neuropsychology*, 1, 83-116.

Milner, B. (1971) Interhemispheric differences in the localisation of psychological processes in man. *British Medical Bulletin*, 27, 272-7.

Miranda, S. B. and Fantz, R. L. (1974) Recognition memory in Down's syndrome and normal infants. *Child Development*, 45, 651–60.

Morton, J., and Patterson, K. (1980) A new attempt at an interpretation, or, an attempt at a new interpretation. In M. Coltheart, J. C. Marshall, and K. Patterson (eds) *Deep Dyslexia*, London: Routledge and Kegan Paul.

Navon, D. (1984) Resources - a theoretical soup stone? *Psychological Review*, 91, 216–34.

Neisser, U. (1983) Components of intelligence or steps in routine procedures? *Cognition*, 15, 189–97.

Netley, C., and Rovet, J. (1982) Verbal deficits in children with 47,XXY and 47,XXX karotypes: a descriptive and experimental study. *Brain and Language*, 17, 58-72.

Nettelbeck, T. (1982) Inspection time: an index for intelligence? *Quarterly Journal of Experimental Psychology*, 34A, 299-312.

Nettelbeck, T. (1985) Inspection time and IQ. Paper presented at the second meeting of the International Society for the Study of Individual Differences, Santa Feliu de Guixols, Catalonia, Spain.

Nettelbeck, T. (1987) Inspection time and intelligence. In P. A. Vernon (ed.), *Speed of Information Processing and Intelligence*, New York: Ablex.

Nettelbeck, T., and Kirby, N. H. (1983) Measures of timed performance and intelligence. *Intelligence*, 7, 39-52.

Nettelbeck, T., and Lally, M. (1976) Inspection time and measured intelligence. *British Journal of Psychology*, 67, 17-22.

Nettelbeck, T., and Lally, M. (1979) Age, intelligence and inspection time. *American Journal of Mental Deficiency*, 83, 398-401.

Nettelbeck, T., and Wilson, C. (1985) A cross sequential analysis of developmental differences in speed of visual information processing. *Journal of Experimental Child Psychology*, 40, 1-22.

Nettelbeck, T., and Young, R. (1989) Inspection time and intelligence in 6-year-old children. *Personality and Individual Differences*, 10, 605-14.

Nettelbeck, T., Hirons, A., and Wilson, C. (1984) Mental retardation, inspection time, and central attentional impairment. *American Journal of Mental Deficiency*, 89, 91-8.

Newcombe, F., and Young, A. W. (1989) Prosopagnosia and object agnosia without covert recognition. *Neuropsychologia*, 27, 179-91.

Nigro, G. N., and Roak, R. M. (1987) Mentally retarded and nonretarded adults' memory for spatial location. *American Journal of Mental Deficiency*, 91, 392-7.

O'Connor, N. (1989) The performance of the 'idiot savant': implicit and explicit. *British Journal of Disorders of Communication*, 24, 1-20.

O'Connor, N., and Hermelin, B. (1981) Intelligence and learning: specific and general handicap. In M. Friedman, J. P. Das, and N. O'Connor (eds), *Intelligence and Learning*, New York: Plenum.

O'Connor, N., and Hermelin, B. (1983) The role of general ability and specific talents in information processing. *British Journal of Developmental Psychology*, 1, 389-403.

O'Connor, N., and Hermelin, B. (1984) Idiot savant calendrical calculators: maths or memory? *Psychological Medicine*, 14, 801-6.

O'Connor, N., and Hermelin, B. (1991) A Specific Linguistic Ability. *American Journal on Mental Retardation*, 95, 673-80.

O'Connor, M., Cohen, S., and Parmelee, A. H. (1984) Infant auditory discrimination in pre-term and full-term infants as a predictor of 5-year intelligence. *Developmental Psychology*, 20, 159-65.

Ornstein, R. E. (1972) *The Psychology of Consciousness*. San Francisco: Freeman.

Paivio, A. (1971) *Imagery and Verbal Processes*. New York: Holt, Rinehart, and Winston.

Pascual-Leone, J. (1970) A mathematical model for the transition rule in Piaget's developmental stages. *Acta Psychologia*, 32, 301-45.

Peck, J. (1987) *The Chomsky Reader*. London: Serpent's Tail.

Pennington, B. F., and Smith, S. D. (1983) Genetic influences on learning disabilities and speech and language disorders. *Child Development*, 54, 369-87.

Piaget, J. (1953) *The Origin of Intelligence in the Child*. London: Routledge and Kegan Paul.

Piaget, J. (1954) *Origins of Intelligence*. New York: Basic Books.

Plomin, R. (1990) *Nature and Nurture: An Introduction to Human Behaviour Genetics*. Pacific Grove, California: Brooks/Cole.

Plomin, R., and Daniels, D. (1987) Why are children in the same family so different from one another? *Behavioral and Brain Sciences*, 10, 1-60.

Posner, M. I., and Mitchell, R. F. (1967) Chronometric analysis of classification. *Psychological Review*, 74, 392-409.

Pylyshyn, Z. W. (1973) What the mind's eye tells the mind's brain: a critique of mental imagery. *Psychological Bulletin*, 80, 1-24.

Pylyshyn, Z. W. (1979) Validating computational models: a critique of Anderson's indeterminacy of representation claim. *Psychological Review*, 86, 383-94.

Pylyshyn, Z. W. (1980) Computation and cognition: issues in the foundations of cognitive science. *Behavioral and Brain Sciences*, 3, 111-69.

Pylyshyn, Z. W. (1984) *Computation and Cognition: Toward a Foundation for Cognitive Science*. Cambridge, Mass.: MIT/ Bradford Books.

Rabbitt, P. M. A. (1985) Oh *g* Dr. Jensen! or, *g*-ing up cognitive psychology? Open peer commentary, *Behavioral and Brain Sciences*, 8, 238-9.

Rimland, B. (1978) Savant abilities of autistic children and their cognitive implications. In G. Serban (ed.), *Cognitive Defects in the Development of Mental Illness*, New York: Brunner/Mazel.

Robinson, D. L., Haier, R., Braden, W., and Krengel, M. (1984) Psychometric intelligence and visual evoked potentials: a replication. *Personality and Individual Differences*, 5, 487-9.

Rose, D. H., Slater, A. S., and Perry, H. (1986) Prediction of childhood intelligence from habituation in early infancy. *Intelligence*, 10, 251-63.

Rovet, J., and Netley, C. (1983) The triple X chromosome syndrome in childhood: recent empirical findings. *Child Development*, 54, 831-45.

Rushton, J. P. (1988) Race differences in behaviour: a review and an evolutionary analysis. *Personality and Individual Differences*, 9, 1009-24.

Rushton, J. P. (1990) Race differences, r/K theory and a reply to Flynn. *The Psychologist: Bulletin of the British Psychological Society*, 3, 195-7.

Sacks, O. (1985) *The Man who Mistook his Wife for a Hat*. London: Duckworth.

Scheerer, M., Rothmann, E., and Goldstein, K. (1945) A case of 'idiot savant': an experimental study of personality organisation. *Psychological Monographs*, 58, 1-63.

Schucard, D. W., and Horn, J. L. (1972) Evoked cortical potentials and measurement of human abilities. *Journal of Comparative and Physiological Psychology*, 78, 59-68.

Shallice, T. (1988) *From Neuropsychology to Mental Structure*. Cambridge: Cambridge University Press.

Shankweiler, D., and Crain, S. (1986) Language mechanisms and reading disorder: a modular approach. *Cognition*, 24, 139-68.

Shannon, C. E., and Weaver, W. (1949) *The Mathematical Theory of Communication*. Urbana, Ill.: University of Illinois Press.

Shepard, R. N. and Cooper, L. A. (1982) *Mental Images and their Transformations*. Cambridge, Mass.: MIT Press.

Shepard, R. N. and Metzler, J. (1971) Mental rotation of three-dimensional objects. *Science*, 171, 701-3.

Shiffrin, R. M., and Schneider, W. (1977) Controlled and automatic human information processing II: Perceptual learning, automatic attending and a general theory. *Psychological Review*, 84, 127–90.

Sloboda, J. A., Hermelin, B., and O'Connor, N. (1985) An exceptional musical memory. *Music Perception*, 3, 155–70.

Smith, G. A., and Stanley, G. (1983) Clocking *g*: relating intelligence and measures of timed performance. *Intelligence*, 7, 353–68.

Smith, S. B. (1983) *The Great Mental Calculators: The Psychology, Methods and Lives of Calculating Prodigies*. New York: Columbia University Press.

Snowling, M., and Frith, U. (1986) Comprehension in 'hyperlexic' readers. *Journal of Experimental Child Psychology*, 42, 392–415.

Spearman, C. (1904) 'General intelligence', objectively determined and measured. *American Journal of Psychology*, 15, 201–93.

Spearman, C. (1927) *The Abilities of Man*. London: Macmillan.

Spelke, E. S. (1987) Where perceiving ends and thinking begins: the apprehension of objects in infancy. In A. Yonas (ed.), *Perceptual Development in Infancy*, Minnesota Symposia on Child Psychology, Hillsdale, N.J.: Lawrence Erlbaum Associates.

Spitz, H. H. (1982) Intellectual extremes, mental age, and the nature of human intelligence. *Merrill-Palmer Quarterly*, 28, 167–92.

Spitz, H. H. (1983) Critique of the developmental position in mental retardation research. *Journal of Special Education*, 17, 261–94.

Spitz, H. H. (1988) Wechsler subtest patterns of mentally retarded groups: relationship to *g* and estimates of heritability. *Intelligence*, 12, 279–97.

Springer, S. P. and Deutsch, G. (1981) *Left Brain, Right Brain*. San Francisco: Freeman.

Stanovich, K. E. (1986) Explaining the variance in reading ability in terms of psychological processes: what have we learned? *Annals of Dyslexia*, 35, 67-95.

Stanovich, K. E., Nathan, R. G., and Zolman, J. E. (1988) The developmental lag hypothesis in reading: longitudinal and matched reading-level comparisons. *Child Development*, 59, 71-86.

Stern, W. (1912) *Die psychologische Methoden der Intelligenzprüfung*. Leipzig: Barth.

Sternberg, R. J. (1981) Novelty-seeking, novelty-finding, and the developmental continuity of intelligence. *Intelligence*, 5, 149–55.

Sternberg, R. J. (1983) Components of human intelligence. *Cognition*, 15, 1-48.

Sternberg, R. J. (1984) Toward a triarchic theory of human intelligence. *Behavioral and Brain Sciences*, 7, 269-315.

Sternberg, R. J. (1985) *Beyond IQ: A Triarchic Theory of Human Intelligence*. Cambridge: Cambridge University Press.

Sternberg, S. (1966) High speed scanning in human memory. *Science*, 153, 652-4.

Sternberg, S. (1969) The discovery of stages: extension of Donder's method. *Acta Psychologica*, 30, 276-315.

Stigler, J. W., Nusbaum, H. C., and Chalip, L. (1988) Developmental differences in speed of processing: central limiting mechanism or skill transfer? *Child Development*, 59, 1144–53.

Temple, C. M. (1991) Procedural dyscalculia and number fact dyscalculia: double dissociation in developmental dyscalculia. *Cognitive Neuropsychology,* 8, 155–76.

Temple, C. M., and Marshall, J. C. (1983) A case study of developmental phonological dyslexia. *British Journal of Psychology,* 74, 517–33.

Terman, L. M. and Oden, M. H. (1947) *The Gifted Child Grows Up: Twenty-Five Years Follow-Up of a Superior Group,* vol. 4 of *Genetic Studies of Genius,* ed. L. M. Terman, Stanford, Calif.: Stanford University Press.

Thal, D., Bates, E., and Bellugi, U. (1989) Language and cognition in two children with Williams syndrome. *Journal of Speech and Hearing Research,* 32, 489–500.

Theobold, T. M., Hay, D. A., and Judge, C. (1987) Individual variation and specific cognitive deficits in Fra (X) syndrome. *American Journal of Medical Genetics,* 28, 1–11.

Thurstone, L. L. (1938) *Primary Mental Abilities.* Chicago: University of Chicago Press.

Tolman, E. C. (1933) Sign-Gestalt or conditioned reflex? *Psychological Review,* 40, 246–55.

Van der Wissel, A., and Zegers, F. E. (1985) Reading retardation revisited. *British Journal of Developmental Psychology,* 3, 3–9.

Vernon, P. A. (1981) Reaction time and intelligence in the mentally retarded. *Intelligence,* 5, 345–55.

Vernon, P. A. (1983) Speed of information processing and intelligence. *Intelligence,* 7, 53–70.

Vernon, P. E. (1950) *The Structure of Human Abilities.* London: Methuen.

Vetterli, C. F., and Furedy, J. J. (1985) Evoked potential correlates of intelligence: some problems with Hendrickson's string measure of evoked potential complexity and error theory of intelligence. *International Journal of Psychophysiology,* 3, 1–3.

Vickers, D. (1970) Evidence for an accumulator model of psychophysical discrimination. *Ergonomics,* 13, 37–58.

Vickers, D., and Smith, P. L. (1986) The rationale for the inspection time index. *Personality and Individual Differences,* 7, 609–23.

Warrington, E. (1982) The fractionation of arithmetic skills: a single case study. *Quarterly Journal of Experimental Psychology,* 34A, 31–51.

Warrington, E., and McCarthy, R. (1987) Categories of knowledge: further fractionation and an attempted integration. *Brain,* 110, 1273–96.

Warrington, E., and Shallice, T. (1984) Category specific semantic impairments. *Brain,* 107, 829–53.

Weiskrantz, L. (ed.) (1988) *Thought without Language.* Oxford: Oxford Scientific Publications.

Weiss, B., Weisz, J. R., and Bromfield, R. (1986) Performance of retarded and non-retarded persons on information processing tasks: further tests of the similar structure hypothesis. *Psychological Bulletin,* 100, 157–75.

Weisz, J. R. and Yeates, K. O. (1981) Cognitive development in retarded and non-retarded persons: Piagetian tests of the similar structure hypothesis.

Psychological Bulletin, 90, 153–78.

Weisz, J. R. and Zigler, E. (1979) Cognitive development in retarded and non-retarded persons: Piagetian tests of the similar sequence hypothesis. *Psychological Bulletin*, 86, 831–51.

Wetherford, M. J. and Cohen, L. B. (1973) Developmental changes in infant visual preferences for novelty and familiarity. *Child Development*, 44, 416–24.

Wickens, C. D. (1974) Temporal limits on human information processing: a developmental study. *Psychological Bulletin*, 81, 739–55.

Wilson, C., and Nettelbeck, T. (1986) Inspection time and the mental age deviation hypothesis. *Personality and Individual Differences*, 7, 669–75.

Wilson, E. O. (1975) *Sociobiology: The New Synthesis*. Cambridge, Mass.: Harvard University Press.

Yule, W. (1985) Comments on van der Wissel & Zegers: reading retardation revisited. *British Journal of Developmental Psychology*, 3, 11-13.

Yule, W., Gold, R. D., and Busch, C. (1982) Long-term predictive validity of the WPPSI: an 11-year follow-up study. *Personality and Individual Differences*, 3, 65-71.

Zigler, E., and Balla, D. (1977) *Mental Retardation: The Developmental–Difference Controversy*. Hillsdale, N. J.: Lawrence Erlbaum Associates.

Zuckerman, M., and Brody, N. (1988) Oysters, rabbits and people: a critique of 'Race differences in behaviour' by J. P. Rushton. *Personality and Individual Differences*, 9, 1025–33.

Index of Names

Note: Page references in italics indicate figures

Aitken, Alexander 181
Anderson, J. R. 92, 213
Anderson, M. 47, 160–6, 175, 178
Anderson, M. *et al.* 182–3
Annett, M. and Manning, M. 85

Baddeley, A. 136, 190
Baillargeon, R. 75
Baron-Cohen, S. *et al.* 10, 64, 130, 195
Bayley, N. 121, 123
Binet, A. 6, 19–22, 27
Bishop, D. V. M. 85, 174
Blinkhorn, S. F. and Hendrickson, D. E. 51
Bornstein, M. H. and Sigman, M. D. 123–4
Brand, C. 45
Brand, C. and Deary, I. J. 45
Brooks, J. and Weinraub, M. 122
Bryant, P. E. and Impey, L. 193, 195
Bryant, P. E. and Trabasso, T. 154
Burt, C. 25, 200

Cahan, S. and Cohen, N. 168
Carey, S. 114
Carpenter, P. A. *et al.* 38 n.1
Carroll, J. B. *et al.* 133
Case, R. 114, 135–7, 139, 145–9, 149 n.1, 158, 167
Ceci, S. J. and Liker, J. K. 21

Chi, M. T. H. 145
Chi, M. T. H. and Ceci, S. J. 145
Chomsky, N. 17–18, 63, 72, 74
Clark, H. H. and Chase, W. G. 32, 91
Cohen, L. B. and Gelber, E. R. 126
Colborn, 181–2
Colombo, J. C. *et al.* 125
Cooper, L. A. and Shepard, R. N. 155
Corballis, M. C. 84
Courchesne, E. *et al.* 50
Cromer, R. F. 72, 120, 174–5

Das, J. P. *et al.* 83
Das Gupta, P. and Anderson, M. 75
Deary, I. J. and Pagliari, C. 198
Dempster, F. N. 145
Dennis, M. 83
De Renzi, E. 71
Detterman, D. K. 41, 199
Detterman, D. K. and Daniel, M. H. 29, 101, 199
Donchin, E. and Coles, M. G. H. 51
Donders 36

Egan, V. 50
Elliott, C. D. 20, 26
Ellis, A. W. 192–3
Ellis, A. W. and Young, A. W. 70
Ellis, H. D. *et al.* 71 n.5

Index of Subjects

Note: Page references in italics indicate figures

Index compiled by Meg Davies
(Society of Indexers)